Structural Classification of Minerals

Volume I

Solid Earth Science Library

Volume 11

Structural Classification of Minerals

Volume I: Minerals with A, $A_m B_n$ and $A_pB_qC_r$ General Chemical Formulas

by

J. LIMA-DE-FARIA
Centro de Cristalografia e Mineralogia,
Instituto de Investigação Científica Tropical, Lisbon, Portugal

KLUWER ACADEMIC PUBLISHERS
DORDRECHT / BOSTON / LONDON

Library of Congress Cataloging-in-Publication Data is available.

ISBN 978-90-481-5680-1

Published by Kluwer Academic Publishers,
PO Box 17, 3300 AA Dordrecht, The Netherlands.

Sold and distributed in North, Central and South America
by Kluwer Academic Publishers,
101 Philip Drive, Norwell, MA 02061, U.S.A.

In all other countries, sold and distributed
by Kluwer Academic Publishers,
P.O. Box 322, 3300 AH Dordrecht, The Netherlands.

Printed on acid-free paper

Cover illustration: Packing drawing of a possible binary compound AB (Barlow, 1898, Fig. 8, p. 453); today known to correspond to halite, NaCl

To my family

Classification, as the author insists, has been proved to be not only an important but an integral part of the scientific enterprise.

Kathleen Lonsdale[a]

A good classification is a research tool.
It is the base of any theoretical work.

Jean Orcel[b]

To change is always seeming fickleness.
But not to change with the advance of science is worse;
it is persistence in error....

James Dwight Dana[c]

[a] In "Origins of Science of Crystals". *Chemistry and Industry*, 1967, 57–58 (p. 57).

[b] In "Essai sur le concept d'éspèce et les classifications en Mineralogie et Petrographie", *Mineralogie, Cristallographie, Aspects actuels. Soc. Miner. Crist. de France*, 1954, 397–432 (p. 404).

[c] In the Preface of his "*System of Mineralogy*" (third edition). John Wiley, New York, 1850.

Contents

Foreword

In his foreword to *Structural Mineralogy. An Introduction* (Lima-de-Faria, 1994) P.B. Moore emphasized that this book "is really not an end in itself. Rather it is a rallying call to urge further clarification, representation and systematization of already known structures". If we consider the new book by Lima-de-Faria, *Structural Classification of Minerals*, in this context, we can ask what kind of new mineralogical data it contains.

The twentieth century was characterized by great progress in the study of minerals. Less than 100 minerals were known up until 1800. Since that time, the rate of discovery of new minerals is steadily increasing. Now it is found that natural processes select some 4000 mineral species, and this number is increasing by 50–60 minerals every year. These data extend the scientific ideas about the forms of concentration of chemical elements, about transformations of minerals during different processes, about classifications of minerals, etc.

A classification is one of the fundamental aspects of modern mineralogy. Facts and phenomena must be put in order before we are able to understand them. Classification in mineralogy provokes one's imagination and helps to discover the parameters which control the formation of definite structural types.

It is well known that the classification of minerals has changed throughout the ages. Its criteria followed the development of mineralogical science. The main criterion was first based on practical purposes, then on physical properties, later on chemical properties. Mineral classification today is largely structural, where the relation and hierarchy between minerals are based on structure similarity. This approach began to develop after 1913, when the first structures of minerals were determined, and the structural criterion for classification was taken into account. The first classification of this type, which takes into consideration the distribution of bonds in a structure, was that of silicates proposed by Machatschki (1928) and developed by Bragg (1930) and Naray-Szabo (1930).

The pure structural classification of minerals was first proposed by J. Lima-de-Faria in 1983. It corresponds to the application of the general structural classification of inorganic compounds (Lima-de-Faria & Figueiredo, 1976) to minerals, which are an integral part of them. The most general approach of the structural systematics is based on the analysis of the strength distribution and of the directional character of the bonds in crystal structures.

There are atoms that are more tightly bound, and these assemblages are called structural units. They are considered as the main basis for the structural classification of minerals. Thus there are five main categories of structures: atomic or close-packed, group, chain, sheet and framework, according to their dimensionality. This approach to the analysis of the crystal structures was approved by the IUCr Commission on Crystallographic Nomenclature in 1990. Later on, in 1994, J. Lima-de-Faria applied the structural classification to the most common minerals (about 500 minerals organized into 230 structure types). Now his task has become much more ambitious: to cover approximately 3500 mineral species.

I am sure that some mineralogists will be confused when they discover in the same section of the classification chemically different minerals such as periclase, halite, galena, osbornite, etc. One can agree or disagree with such an approach, however everybody should accept that there are clear and logical principles in the system of minerals proposed by José Lima-de-Faria. Several chapters of his new book explain many crystal chemical terms and phenomena, which were not defined clearly enough in earlier publications and which are used by mineralogists in their everyday work. It includes structural notation, chemical and structural formulas, the mode of presentation of crystal structures, the correlation between crystalline structure and properties of minerals, and some other problems. Consequently this work is of particular interest to teachers, students and researchers in crystallography, mineralogy and inorganic crystal chemistry in academia.

Sir James Barrie (1860–1937) once noted that "a man of science appears to be the only man who has something to say – and the only man who does not know how to say it". This new book written by José Lima-de-Faria justifies only the first part of this statement and proves that the author really knows how to explain his ideas about the mysterious world of minerals.

D. Yu. Pushcharovsky
Professor at Moscow State University

Preface and Acknowledgements

Now that the structural classification of minerals has been proposed and schematized (Lima-de-Faria, 1983) and applied to the most common minerals (Lima-de-Faria, 1994), it is necessary to extend it to the whole domain of minerals (approximately 3800). As Fedorov said in 1913 "the structural classification of minerals is the natural classification".

Such a large project has to be undertaken in several stages, and we chose to tackle this problem by ordering the minerals from the simple to the more complex general chemical formulas, like Machatschki did in 1953. Consequently this first part is concerned with minerals of general chemical formulas: A, A_mB_n, and $A_pB_qC_r$, involving 960 mineral species. Of these, 922 correspond to this first part, 2 are amorphous, and 36 have structural formulas of the type $A_pB_qC_rD_s$, and consequently will not be included in this first part of our study.

We want to give credit to the many mineralogists and crystallographers who, since 1913, have contributed to the determination of mineral structures. Their work has, in many cases, remained of very little use. It is now time to awake and do it justice.

Grateful acknowledgement is made of the support facilities afforded by the Instituto de Investigação Científica Tropical, in particular by its President, Professor Joaquim Cruz e Silva, and by the Director of the Crystallographic and Mineralogical Center, my colleague Professor Maria Ondina Figueiredo. Thanks are due to my wife Natasha for assistance in the computer work, and to my friend Arnaldo Silvério for helping with the English. Professor Dmitry Pushcharovsky, besides having agreed to write the Foreword, has encouraged this work from the beginning, and I am very much indebted for his collaboration.

I had to ask permission of some authors and publishers to reproduce the figures and tables. I am grateful to Professor F.C. Hawthorne and to the following publishers: American Mineralogist, Plenum Publishing Corporation, Mineralogical Society of America, and the Institut für Geschichte der Naturwissenschaften.

We are convinced that the structural classification of minerals will open new avenues in the development of mineralogical science. The relationship between structure and properties has to be improved, in order that we may take more advantages of this classification.

The present situation is similar to that at the time of James Dwight Dana (1850) when he proposed the general use of the chemical classification, instead of the physical classification. Like him we have to face many difficulties and oppositions. Let us hope that the new structural classification will not take too long a time to be understood and widely accepted.

CHAPTER 1

The structural classification of minerals

1.1. Introduction

A first attempt to present a general classification of minerals on structural grounds and in a systematic way was made by the author in 1983. Hawthorne in 1984 and 1985 suggested a structural classification of minerals based on the polymerization of the coordination polyhedra, as a consequence of the application of the bond-valence theory to inorganic structures. In 1994 the author applied the structural classification to the most common minerals (approximately 500 mineral species). Our aim now is to develop a complete structural classification of all minerals.

However, in order to deal with the whole domain of minerals (approximately 3800 species) we decided to divide this work in three main parts, and follow the subdivisions used by Machatschki (1953) in his book *Spezielle Mineralogie*. This division and corresponding subdivisions are the following:

Minerals without water molecules in their structure	Part I –	A A_mB_n $A_pB_qC_r$
	Part II –	$A_pB_qC_rD_s$ $A_pB_qC_rD_sE_x$ $A_pB_qC_rD_sE_xF_y$ $A_pB_qC_rD_sE_xF_yG_z$ $A_pB_qC_rD_sE_xF_yG_z\ldots$
Minerals with water molecules in their structure	Part III –	$A_pB_q\ldots E_xF_y\ldots n$ aq.

For the enumeration of the several mineral species which are now known, and considered valid, we have used an up-to-date list of minerals from the book by Nickel and Nichols (1991), the *Mineral Reference Manual*. This book, which lists references where important structural data can be found for each mineral, is all the more appealing since one of its authors is Vice-President of the Commission on Mineral Names and New Minerals (CNMMN).

In the above-mentioned book Nickel and Nichols consider five kinds of mineral species: **G** generally accepted, **A** approved by the CNMMN, **P** polytypes, **Q** questionable, and **D** discredited. It is clear that we have not included in our study the kinds **Q** and **D**.

The spelling of the mineral names and the chemical composition of the various mineral species were also taken from this book.The transformation of the chemical into general formulas A, AB, ..., obeyed certain rules. For instance, the chemical elements separated by commas and enclosed within curved brackets, like (Ta,Nb), are considered as a sole A atom (that is, only one atomic structural position). Certain groups of atoms, like OH or NH_4, were also represented as a sole atom A. However (A_m,B_n) corresponds to A_{m+n}.

Due to the pioneering character of the structural classification and the consequent difficulties encountered, apart from the *classified minerals*, we have considered two other categories: minerals *tentatively classified*, where the structural data could not be interpreted in all their details, and *not yet classified*, including those for which the structure is not yet known, or those whose structure has been determined but which we were not able to interpret (the latter are marked with the symbol ●).

1.2. The structural classification of minerals has to fit the general structural classification of inorganic compounds

Minerals are part of the inorganic compounds and therefore the structural classification of minerals cannot be different from that of the inorganic compounds. The structural classification of minerals must also not contradict the structural classification of silicates, because this has proved so fruitful that its structural principles should be preserved. Consequently, the classifications of the silicates, of minerals, and of inorganic compounds are intimately related.

If we look now at the classification of crystals in general (Table I), we find that the proposed general criterion of classification also corresponds to the extension of the structural classification of silicates to the whole domain of crystallography. Table I lists some of the main attempts to organize a general classification of crystals; they were rather similar, differing basically in the first category: coordination or atomic (close-packed). Hawthorne (1983) has proposed a general classification of crystals based on the polymerization of coordination polyhedra with higher bond valences. The classification developed by Lima-de-Faria and Figueiredo (1976) can also be regarded as general.

Regarding the chemical classification, there is a common perspective of the various authors, with only slight differences. The chemical classifications of minerals of Strunz (1941), Povarennykh (1966), Liebau (1962) or Zoltai (1960) differ very little. However, one could imagine that several structural classifications of minerals should exist, according to the purpose and the group of minerals under consideration. This situation would complicate very much the development of mineralogy. As with the chemical classification, a sole kind of structural classification must be devised, possibly with slight variants.

1.3. The scheme of the structural classification of minerals

As we mentioned before, the structural classification of minerals corresponds mainly to an extension of the structural classification of silicates to the whole domain of minerals. There are however some differences. Instead of the four categories considered up to now for silicates, we have established five structural categories of minerals. It is well known that the so-called polymerization process of silica tetrahedra starts with a sole tetrahedron, proceeds to assemblages of tetrahedra forming finite groups, then infinite chains of tetrahedra, infinite sheets, and finally infinite three-dimensional frameworks. This scheme appears to be incomplete. The polymerization process should start with individual atoms, which polymerize into finite groups (tetrahedra, octahedra, rings of tetrahedra, etc.), those in turn into infinite chains, infinite sheets and finally into frameworks. The new categories of minerals are therefore based on these five main structural units:

Close-packed (individual atoms are the structural units forming the close packing)
Group (finite groups, finite chains, rings, cages, etc.)
Chain (infinite chains)
Sheet (infinite sheets)
Framework (three-dimensional infinite network)

Table I. General classification of crystal structures

Laves (1930)	Niggli (1945)	Garrido and Orland (1946)	Bokii (1954)	De Jong (1959)	Lima-de-Faria and Figueiredo (1976)
	Isolated particules		Coordination	Coordination	Atomic (or close-packed)
Islands	Molecules Radicals	Isolated groups	Islands	Molecules Radicals	Groups
Chains	Chains	Chains	Chains	Chains	Chains
Sheets	Sheets	Sheets	Sheets	Sheets	Sheets
Frameworks	Frameworks	Frameworks	Frameworks	Frameworks	Frameworks

For instance, halite has been considered a coordination structure (Bokii, 1960, p. 193, De Jong, 1959, p. 144) but we think that it should be classified as a close-packed mineral . In fact the Cl atoms form a cubic closest packing with Na atoms occupying all the octahedral voids.

Some of the silicates that were classified as group minerals based on isolated tetrahedra are here classified as close-packed. This is the case, for instance, of forsterite, Mg_2SiO_4, which is considered a hexagonal closest packing of the oxygens with magnesium in octahedral voids and silicon in tetrahedral holes, that is $Mg_2^o\ Si\ ^t[O_4]^h$, according to Bragg (1929) and Megaw (1973). There is a spinel-like form of olivine, whose structural formula is $Mg_2^oSi^t\ [O_4]^c$; no doubt it should be considered a close-packed mineral, rather than a group.

The structural classification of minerals, like that of the silicates, is based on the bond strength distribution and on the directional character of the bonds. According to the Laves (1956) principles of structure stability (highest density, highest symmetry and maximum coordination), when there are no directional bonds in the structure the mineral will tend to be as dense and symmetrical as possible, and will be a close-packed mineral. However some difficulties can arise in the application of these principles. For instance, $Cd^o[I_2]^h$, which is a close-packed mineral, has been sometimes regarded as a sheet structure due to its good cleavage along a certain plane. This stems from the use of the bond strength distribution alone without taking into consideration the non-directional character of the bonds between the iodine atoms. This structure is better described as an hexagonal closest packing of the iodine atoms with cadmium atoms occupying certain octahedral voids. The distribution of the cadmium in layers is just one of the possible distributions for ½ as the proportion of octahedral occupation in hexagonal closest packing. Whenever a structure is based on a close packing of the larger atoms with smaller atoms occupying its interstices the mineral should be classified as close-packed. Due to the directional character of the bonds or to a special coordination of certain atoms, the close packing may be deformed, becoming less dense, with large voids, giving rise to high structural units, the mineral being no longer a close-packed, but an open structure, of higher category.

The five main categories of minerals can be subdivided according to their complexity (Lima-de-Faria, 1994). The close-packed minerals are called *layered* if they can be imagined as formed by the stacking of approximately plane layers, and *non-layered* if the packing is three-dimensional, not being subdivided in layers, but with a very high packing efficiency. An example is garnet with a high packing efficiency, similar to many layered close-packed minerals.

When a mineral is built up of several structural units of different dimensionalities it is classified according to the structural unit of highest dimensionality.

1.4. The structural notation; chemical and structural formulas, the symbol of the structure type

For the structural notation we use structural formulas, and to define the structure types we use the structural formula of the so-called aristotype (in the sense of Megaw, 1973), that is, the compound of higher symmetry and more commonly used to express a certain atomic arrangement.

The structural formula of a mineral is just its chemical formula with the structural information added on in the simplest and most self-explanatory way. We add to the chemical formula the main structural characteristics, that is, the coordination of the atoms, the indication of the structural units, their dimensionality and, whenever possible, the way they are packed together.

If we consider, for instance, the mineral halite, the chemical formula is NaCl. The coordination of the Na atoms is octahedral and the Cl atoms form a cubic closest packing. Therefore the structural formula is $Na^o[Cl]^c$, where the superscript o means octahedral, and the structural units are the individual atoms of Cl, which are written within normal square brackets, forming the cubic closest packing c.

Consequently, for halite,

NaCl	$Na^o[Cl]^c$
chemical formula	structural formula

The general notation for *the coordination* of the atoms is the coordination number, *as a superscript, within small square brackets*. However in the case of more common coordinations we use lower-case letters, as self-explanatory as possible. Examples are: o for octahedral coordination, t for tetrahedral, cb for cubic, etc.

The structural units are always written *within normal square brackets*.

The dimensionality of the structural units is expressed *within curly brackets* with the aid of the symbol infinity, ∞, for the structural units with infinite dimension like the chains, sheets and frameworks, for which we use the symbols {1∞}, {2∞}, and {3∞}, respectively. They correspond to the Machatschki (1947) symbols, only written in a computer-friendly way. For finite dimensionality instead of zero infinity we use the symbol {g} for any kind of groups, and no symbol for the individual atoms of the close packings. If the mineral is a chain, a sheet or a framework, and has no interstitial atoms and therefore the structural formula starts with the dimensionality symbol, then we do not use the curly brackets, because there is no reason for misunderstanding.

The packing of the structural units is given as superscript, using letters as self-explanatory as possible, and *within angle brackets* (Lima-de-Faria et al., 1990). It is formed by the symbol of the layer followed by the symbol of the stacking. For instance, the closest-packed layer is designated by T (for triangular tesselation) and the cubic closest stacking by c, therefore the notation for the cubic closest-packing is <Tc>. The hexagonal closest packing is <Th>, the simple hexagonal packing is <Ts> (s for superimposition of the layers), etc. However when the symbols are very common, like c, h, or b (the cubic body centred packing), the angular brackets are omitted. For a mineral with general chemical formula $A_pB_qC_rD_s$ a possible structural formula could be

$$A_p^{[m]}B_q^{[n]}\{\}[C_r^{[\gamma]}D_s^{[\delta]}]^{<>}$$

An arrangement of atoms which is geometrically similar to a close packing when considered alone and not linked to higher structural units is written within double angle brackets «». An example is calcite $Ca^{ap}\{g\}[C^{tr}O_3^{«h»}]^c$, where the oxygens may be imagined to form a hexagonal closest packing.

More complex layers are the N21 and the R21. N21 is formed by interconnected triangles and squares, with tetragonal symmetry, in the proportion of two triangles to one square. An example is anyuiite $Au^{acb}\ [Pb_2]^{N21\ f}$, f meaning the kind of stacking. R21 is formed by rows of triangles and rows of squares, also in the proportion 2 : 1. An example is andalusite $Al^o\ Al^{[5]}\ Si^t\ [O_5\ \square]^{R21\ my}$ (see Lima-de-Faria, 1994, pp. 29, 144 and 147).

Maldonite is isotypic with $MgCu_2$ (a Laves phase) and is formed by the heterogeneous packing of layers $\dot{\tilde{T}}$ and K (kagomé layer), and the corresponding structural formula is $[\ [Au_2]\ [Bi]\ ]^{(.\sim TK)\ c}$. The symbol $\dot{\tilde{T}}$ is written .~T due to computer difficulties.

Some crystal structures like stibnite, Sb_2S_3, are a mosaic of linked close-packed blocks (Hellner, 1958). They are called recombination structures (Lima-de-Faria et al., 1990) and the corresponding structural formula is $Sb_2^{[3+4]}\ [S_3]^{\#}$, where # means linked close-packed blocks, or *recombination packing* (previously, this kind of packing was depicted by c¦h). However, this notation does not describe the way the blocks are linked together and needs some improvement, possibly to be based on the work on sulphosalts by Makovicky (1985, 1997a, 1997b).

The frameworks can be imagined as decomposed in special layers, the so-called connected units, which rather than individual layers are layers strongly linked to the rest of the structure. An example is diamond, C, where the connected units are puckered hexagonal layers ~H of carbon atoms; the corresponding structural formula is $3\infty[C^t]^{\sim Hc}$.

When in close-packed minerals the layer description is not applicable, because the packing is three-dimensional, like in garnet, $Ca_3^{do}Al_2^oSi_3^t[O_{12}]^*$, such three-dimensional packing is depicted by an asterisk *. In fact the garnet structure has a high packing coefficient of approximately 60%, which is comparable to that of forsterite, $Mg_2^oSi^t[O_4]^h$, with 70%, and fosterite is certainly based on a hexagonal closest packing of the oxygens.

The other kinds of structural units also tend to pack together as closely as possible. When the structural units are groups, they tend to behave as organic molecules. The molecules stack in such a way that the bumps in one are inserted in the hollows of another molecule. A very complete study of the packing of finite molecules, infinite molecular chains, and infinite molecular sheets in organic structures has been made by Kitaigorodskii (1955). This study may be applied to the corresponding categories of inorganic structural units.

The symbol of the structure type, as mentioned above, corresponds to the structural formula of the so-called aristotype, in the sense of Megaw (1973, p.73) (see also Lima-de-Faria, 1994, p.75).

For *the coordination of the atoms* two levels of symbols have been proposed, namely, complete and simplified. Each complete symbol gives the total number of atoms coordinated to a certain atom, and the type of coordination polyhedron, indicated by lower-case letters. The symbols for the most common coordination polyhedra are presented in Table II.

For the sake of simplification the structural formula includes only the already-mentioned main structural factors. However, it may be useful in certain cases to add more structural information, either to describe more completely the structure, or to enable complicated structures to be compared.

It has also been recognized that when a structure is not close-packed, but belongs to

Table II. Symbols for common coordination polyhedra (adapted from Lima-de-Faria et al., 1990, p. 5)

Coordination polyhedron around atom *A*	Complete symbol	Simplified symbols		
Single neighbour	[1*l*]		[1]	
Two atoms collinear with atom *A*	[2*l*]		[2]	
Two atoms non-collinear with atom *A*	[2*n*]			
Triangle coplanar with atom *A*	[3*l*]		[3]	tr
Triangle non-coplanar with atom *A*	[3*n*]			
Triangular pyramid with atom A in the centre of the base	[4*y*]			
Tetrahedron	[4*t*]	[*t*]	[4]	*t*
Square coplanar with atom *A*	[4l]*	[*s*]		*s***
Square non-coplanar with atom *A*	[4*n*]			
Pentagon coplanar with atom *A*	[5*l*]			
Tetragonal pyramid with atom *A* in the centre of the base	[5*y*]		[5]	
Trigonal bipyramid	[5*by*]			
Octahedron	[6*o*]	[*o*]		*o*
Trigonal prism	[6*p*]	[*p*]	[6]	*p*
Trigonal antiprism	[6*ap*]	[*ap*]		*ap*
Pentagonal bipyramid	[7*by*]		[7]	
Monocapped trigonal prism	[6*p*1*c*]			
Bicapped trigonal prism	[6*p*2*c*]			
Tetragonal prism	[8*p*]			
Tetragonal antiprism	[8*ap*]			
Cube	[8*cb*]	[*cb*]	[8]	*cb*
Anticube	[8*acb*]	[*acb*]		*acb*
Dodecahedron with triangular faces	[8*do*]	[*do*]		*do*
Hexagonal bipyramid	[8*by*]			
Tricapped trigonal prism	[6*p*3*c*]		[9]	
Cuboctahedron	[12*co*]	[*co*]		*co*
Anticuboctahedron (twinned cuboctahedron)	[12*aco*]	[*aco*]		*aco*
Icosahedron	[12*i*]	[*i*]	[12]	*i*
Truncated tetrahedron	[12*tt*]			
Hexagonal prism	[12*p*]			
Frank–Kasper polyhedra with:				
14 vertices	[14*FK*]		[14]	
15 vertices	[15*FK*]		[15]	
16 vertices	[16*FK*]		[16]	

*Also [4*s*]. ** or sq

another structural category, it still can be sliced into layers and, in most cases, the stacking of the layers corresponds to the generalized stackings c or h, that is, ABC or ABA, like the closest packings.

The Subcommittee on Nomenclature of the International Union of Crystallography *has strongly recommended the use of structural formulas*, in order to facilitate the progress of crystal chemistry. Every worker in this field is advised to adhere to such recommendation and to make a point of using structural formulas whenever they can be substituted for chemical formulas.

1.5 Structural derivatives. The measure of the symmetry of crystal structures

Among the relationships that are of especial interest, particular importance should be given to the *structural derivatives*. We shall consider first the derivatives in the sense of Buerger (1947), which include two categories, distortion and substitution derivatives. A *distortion derivative* of a certain structure corresponds to another structure that can be derived from it by a slight distortion. A *substitution derivative* is obtained by replacing one kind of atom by two or more atoms. However there is another category that Buerger did not consider, the *defect derivatives*, which are derived by subtraction of some packing or interstitial atoms, generating vacancies. An example of an ordered defect derivative of perovskite, $Ti^{o}[CaO_3]^{c}$, is dzhalindite, $In^{o}[\square(OH)_3]^{c}$ ($\square$ meaning vacancy), which belongs to the $Re^{o}[\square O_3]^{c}$ inorganic structure type.

With regard to two structures one of which is a distortion derivative of the other, the distortion derivative is the one with less symmetry, and the other is called the basic structure. To decide which is the structure with less symmetry we need to know *how to measure the symmetry of a crystal structure*. Attempts to measure the symmetry of a structure have been based on group–subgroup relations, but this has led to contradictory results. A definition of the measure of the symmetry of a crystal structure was proposed by Lima-de-Faria (1988); it corresponds to *the multiplicity of the general position of the space group*. This definition is the natural extension to the space groups of the one adopted for point groups. In fact the multiplicity of the general form in point groups corresponds to the multiplicity of the general position in space groups. For further details of this definition see Lima-de-Faria (1991, 1994).

When comparing structures which are isopointal structures, with a view to assembling them in structure types, we have first to consider limits of the variability of their parameters that do not change the coordination of the atoms. According to Bergerhoff et al. (1999), to each pair of isopointal structures there corresponds a value of the difference of their coordinates which they called Δ. This value will determine if the structures should be considered isoconfigurational or not, and it will give a measure of their similarity.

1.6. The representation of crystal structures

There are several ways of representing crystal structures (Lima-de-Faria, 1994), but the most common one is the projection along a crystallographic axis. Moore (1992) considers that the crystal representation is very important for the structural interpretation and claims that by choosing the appropriate description he has found many structures based on close packings that previously had not been described as such (Moore, 1995).

This tendency to represent crystal structures on the basis of projections along the crystallographic axes may render the structural understanding difficult, because, in many cases, these axes have no structural interest.

In fact many structures have been found to be close-packed structures. For instance, the structure of hodgkinsonite, $MnZn_2SiO_4(OH)_2$, in projection along the crystallographic axes does not disclose that it is based on a slightly distorted hexagonal closest packing of the oxygens and hydroxyls with Mn in octahedral voids and Zn and Si in tetrahedral voids. The corresponding structural formula is $Mn^{o}Zn_2^{t}Si^{t}[O_4(OH)_2]^{h}$. This was recognized by Rentzeperis, in 1963, and further described in terms of a condensed model by Figueiredo (1976), the structure being built in equal layers (see Lima-de-Faria, 1994, p. 134).

Another curious example is talbochite, $CuCl_2$. Wyckoff, who was a crystallographer acquainted with and very interested in close packings, has not noticed that talbochite is a close-packed mineral (1963, vol. 1, p. 345–346). The same happened with Burns and Hawthorne (1993). This structure is in fact based on a slightly distorted cubic closest packing of the Cl atoms, with Cu atoms occupying octahedral voids in alternate layers, and should be considered as a distortion derivative of cadmium chloride, $Cd^o[Cl_2]^c$. These structures correspond to the same atomic arrangement, but have been considered different because they were represented under different directions (Figure 1). A sphere packing model of cadmium chloride, and the use of the ICSD (Inorganic Crystal Structures Database) and the "Diamond" computer programs, organized by Bergerhoff (1995, 1997), have helped to confirm this relationship. In this typical example, because distortion lowers the symmetry, the corresponding projection along the crystallographic axes masks the real structural nature of the compound.

1.7. The crystalline structure and properties

It is clear that chemical composition is not the most suitable tool to characterize a mineral. In fact, should we merely be told that diamond and graphite are both plain carbon we would not hazard a guess at their properties. However, if we learn that diamond is a framework and graphite a sheet structure, we may readily endow them with quite different properties: diamond is probably a harder mineral, with imperfect cleavage and isometric habit, and graphite a less hard mineral, with perfect cleavage and tabular habit.

There is an intimate relation between the internal structure of minerals and their properties. Povarennykh (1972, p. 104) said: "Polymorphic species (diamond and graphite, calcite and aragonite, kyanite and sillimanite) provide the best evidence that the structure type and the bond distribution are the decisive factors as regards the properties [...]; only a consideration of the structural features (bond distribution and strength) can give a full conception of all the major properties". Povarennyhk summarized these relationships as shown in Table III.

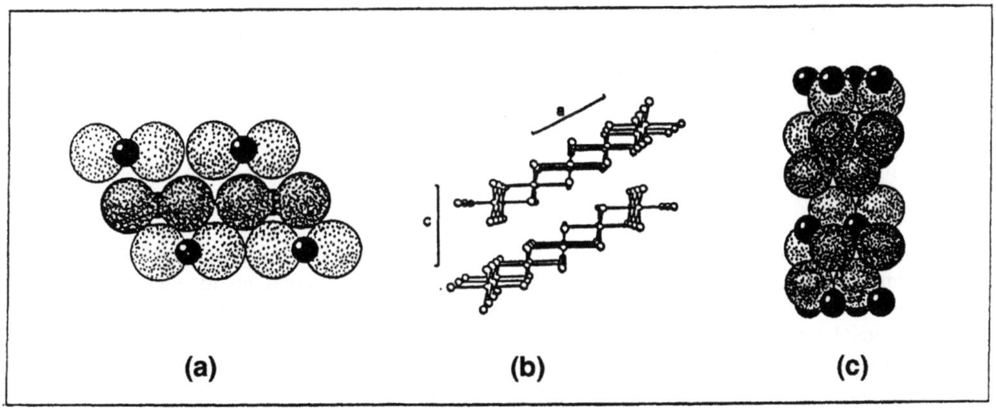

Fig. 1. Structural representation of the talbochite structure, $Cu^o[Cl_2]^c$, C2/m: (a) after Wyckoff (1963, p. 346); (b) after Burns and Hawthorne (1993, p. 188); (c) packing representation of the structure of cadmium chloride, $Cd^o[Cl_2]^c$, R3m, after Wyckoff (1963, p. 271). These three representations correspond to the same structural arrangement; tolbachite is just a distortion derivative of $Cd^o[Cl_2]^c$.

Table III. Relation between crystal structure and properties (after Povarennykh, 1972, p. 109)

Pattern	Index of refraction	Birefringence and optical sign	Cleavage	Hardness	Density	Habit
Coordination	Varies widely with crystallo-chemical factors	None or low (+ and –)	Moderate to perfect in three or more directions	Varies widely with crystallo-chemical factors	Varies widely medium to high	Isometric
Framework	Lowest	Low (+ and –)	Imperfect or moderate in several directions	Medium	Least	Mainly isometric
Ring	Low (inversely related to ring diameter)	Low (mostly –)	Usually imperfect to moderate in one or more directions	Medium to high	Low (inversely related to ring diameter	Mainly isometric (to columnar)
Insular	High (largest relative to other mineral in the class)	Medium, except for minerals with planar radicals (+ and –)	Varies, often moderate or imperfect in several directions	High (largest relative to other minerals in the class)	Usually high (except hydrates)	Mainly isometric
Chain	Medium	High (mostly +)	Moderate to perfect in two directions (occasionally one) parallel to chains	Medium	Medium	Columnar, spiky fibroid, needle
Layer	Medium to low	Very high (mostly –)	Perfect to highly perfect in one direction parallel to layers	Low (increases with strength of bonds between layers	Low (varies with composition	Tabular, platy, foliated, scaly

Although there are many problems, such as polymorphism, diadochy, phase transformations, cleavage, habit and twinning, which can only be clearly interpreted by taking into account the structural characteristics, the question of the relationship between structure and physico-chemical properties is very little developed. According to Megaw (1973, p. viii), " knowledge in this field is still very scanty (except for the very simple structures)".

Certain physical properties are more sensitive to the geometrical arrangement of the atoms in space, that is, to symmetry; for example, elasticity, thermoconductivity, electroconductivity and thermoexpansion. Others, such as cleavage, habit, hardness and twinning, are more related to the kind of structural units and their packing.

Consequently, we are convinced that one of the main negative factors with respect to this problem has been the lack of use of a structural classification and notation – a further important reason for developing the structural classification of minerals.

1.8. The structural classification of minerals is a natural classification

The structural classification of minerals is a natural classification because it is based on the internal structure of minerals which is responsible for their entire properties. It is clear that it is not the chemistry alone that determines the properties of a mineral; as already stated, diamond and graphite have the same chemical composition, but present very different properties because they correspond to different structures.

The natural classification of animals and plants which was developed in 1735 by Linnaeus was based on their internal structure, that is on their functional organs. Scientists could cut up and open the animals and plants to study their anatomy, but at that time the internal structure of minerals could not be oberved or determined (Figure 2). One could only break the minerals into pieces, or observe their surface under a high magnification, to notice only defects and more cracks. The classification of minerals has changed throughout the ages, and each change was always a step further in the direction of their internal structure. Whewell, professor at Cambridge, U.K., had this perception, and stated in 1837: "We cannot get rid of the fundamental conviction that the elementary composition of bodies, since it fixes their essence, must determine their properties [...]. We may begin with the outside, but it is only in order to reach the internal structure".

After the discovery of the application of the X-rays to the determination of crystal structures by Bragg (1913), contact with the internal structure of minerals became possible, and Fedorov (1913) could write: "Only very recently has the principle of crystallochemical analysis become capable of leading to a classification in which every mineral has a strictly defined place. Such a classification cannot be called artificial since its basis is the structure of the mineral, i.e., that which essentially characterizes the nature of each". However the structural classification of minerals could not be used immediately due to the lack of structural information at the time, because only a few structures of minerals were known, which did not enable the generality of such classification. Only nowadays, when most of the mineral structures have been determined, can the structural classification really be developed.

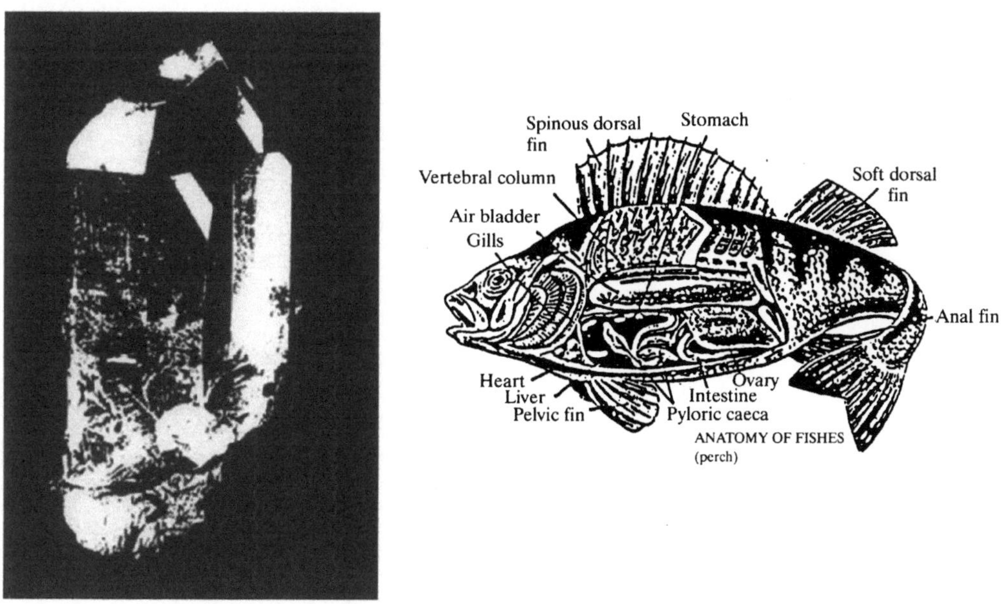

Fig. 2. Quartz crystal and anatomy of a perch (after Lima-de-Faria, 1998, p. 23)

1.9. The similarity between the time of James Dwight Dana and the present

We are facing now very similar difficulties to those found by James Dwight Dana, a hundred and fifty years ago, when he proposed to change from the physical classification of minerals to the chemical classification.

The first main difficulty is concerned with the change of scientific mental habits. There is always a strong reaction against new ideas. Dana in 1850 wrote in the preface to his book: "To change is always seeming fickleness. But not to change with the advance of science is worse; it is persistence in error..."

One of the arguments against the chemical classification was the fact that it grouped together minerals with very different lustres, like galena and blende. Dana replied in his preface to the 1854 edition: "The mind uneducated in Science may revolt at seeing a metallic mineral, as galena, side by side with one unmetallic lustre, as blende; and some systems, in accordance with this prejudice, place these species in separate orders. [...] But the distinction of 'useful' and 'useless', or 'ores' and 'stones', although bearing on 'economy', is not Science". The same happens today with the structural classification. In 1984 Gottardi wrote: "[...] a purely structural classification is unavoidable nowadays, but with the disadvantage of having galena and rocksalt in the same box."

This supports the statement above that, at present, structural classification is facing a number of difficulties as did chemical classification in Dana's time. The parallelism is very close, but with the deeper historical knowledge of today's scientists we hope that such difficulties will soon be overcome.

CHAPTER 2

Presentation of the data

2.1. Selection and presentation of the mineral data; systematic tables

The selection of the minerals known at present (approximately 3800) was made on the basis of the book by Nickel and Nichols, *Mineral Reference Manual*, as already stated.

The systematic description of the structural data of minerals is organized into tables corresponding to the same general chemical formula, and in each of them the minerals are presented in alphabetical order. The tables contain the name, the chemical formula (according to Nickel and Nichols), the corresponding structural formula (whenever possible), the crystal system, space group, the unit cell parameters, the equivalent positions occupied by the atoms, the structure type to which it belongs, and the references which yielded all the data. These tables are ordered from the simple to the more complex general chemical formulas.

The number Z of chemical formulas per unit cell is referred to the given chemical formula and not to the structural formula. The equivalent positions are presented in abbreviated form, according to Wyckoff's nomenclature. Instead of the designation of the rhombohedral system, as used by Nickel and Nichols, we use the trigonal system.

For the selection of the structural data, apart from the references given by Nickel and Nichols, we have used several important books such as those by Wyckoff (1963–1968), Povarennykh (1972), and Strunz (1982).

For each general chemical formula we selected the corresponding structure types, followed by their population, and their derivatives. Minerals which were only tentatively classified are listed separately; they are ascribed a probable atomic arrangement but not a complete detailed structure of the mineral. Minerals not yet classified are also listed separately; they are minerals whose structure is not yet known, or could not be found, or else has been determined but is difficult to interpret in terms of the structural classification (they are marked with the symbol ●).

Table 1

A

NAME	CHEMICAL FORMULA	STRUCTURAL FORMULA	SPACE GROUP	UNIT CELL DIMENSIONS	EQUIVALENT POSITIONS	STRUCTURE TYPE	REFERENCES
ALLARGENTUM	$Ag_{1-x}Sb_x$ ε-form	$[Ag_{0.86},Sb_{0.14}]^n$	Hex.	a=2.95Å c=4.77Å Z=2			Can. Min.,1970,10,163-172; Pov. 205;Am.Min.,1970,56, 638(Abs.).
ALUMINIUM	Al	$[Al]^c$	Cub. $Fm\bar{3}m$	a=4.04Å Z=4	Al(4a)	$[Cu]^c$ COPPER	SB,1,13-14;Am.Min.,1980,65,205 (Abs.);Wyckoff,1,7-10;Hölzel,3.
ANTIMONY	Sb	$2\infty[Sb^{[3n]}]^{-Hc}$	Trig. $R\bar{3}m$	a=4.3084Å a_R=4.506Å c=11.2740Å α=57.110° Z=6 Z_R=2	Sb(6c)	$2\infty[As^{[3n]}]^{-Hc}$ ARSENIC	Acta Cryst.,1963,16,451-453; J.Appl.Cryst.,1969,2,30-36; Pov.,197-198.
ARSENIC	As	$2\infty[As^{[3n]}]^{-Hc}$	Trig. $R\bar{3}m$	a=3.7598Å a_R=4.1320Å c=10.5475Å α=54.126° Z=6 Z_R=2	As(6c)	$2\infty[As^{[3n]}]^{-Hc}$ ARSENIC	J.Appl. Cryst.,1969,2,30-36; Wyckoff,1,32-33;Pov.,191-198; LF,224.
ARSENOLAMPRI-TE	As	$2\infty[As^{[3]}]^{-Hc}$	Orth. Bmab	a=3.63Å Z=8 b=4.45Å c=10.96Å	As(8f)	Dist.deriv. $2\infty[As^{[3n]}]^{-Hc}$ ARSENIC	Am.Min.,1960,45,479-480 (Abs.);SR,42A,23;Wyckoff,1, 30-31.
BISMUTH	Bi	$2\infty[Bi^{[3n]}]^{-Hc}$	Trig. $R\bar{3}m$	a=4.5460 a_R=4.7458Å c=11.862Å α=57.23° Z=6 Z_R=2	Bi(6c)	$2\infty[As^{[3n]}]^{-Hc}$ ARSENIC	J.Appl.Cryst.,1969,2,30-36; Pov.,197-198.
CADMIUM	Cd	$[Cd]^h$	Hex. $P6_3/mmc$	a=2.979Å c/a=1.88 c=5.617Å (c/a ideal=1.63) Z=2	Cd(2c)	Dist.deriv. $[Os]^h$ OSMIUM	Am.Min.,1980,65,1065(Abs.); Bokii,258;Wyckoff,1,8-11; Hölzel,2.
CHAOITE	C		Hex. P6/mmm	a=8.948Å Z=168 c=14.078Å			Am.Min.,1969,54,326(Abs.);Am. Min.,1970,55,1067(Abs.);Hölzel,3.
CHROMFERIDE	$Fe_{1.5}Cr_{0.2}$	$[Fe_{1.5},Cr_{0.2},Y_{0.3}]^b$	Cub. $Pm\bar{3}m$	a=2.859Å Z=1	Cr(2a)		ZVMO,1986,115,355-360;Am. Min.,1988,68,1250(Abs.).
CHROMIUM	Cr	$[Cr]^b$	Cub. $Im\bar{3}m$.	a=2.88Å Z=2		$[Fe]^b$ IRON	Am.Min.,1982,67,854-855 (Abs.);Bokii,239-240;Hölzel,4.
COPPER	Cu	$[Cu]^c$	Cub. $Fm\bar{3}m$	a=3.61496Å (18°C) Z=4	Cu(4a)	$[Cu]^c$ COPPER	Wyckoff,1,7-10;Pov.,193;LF,83.
DIAMOND	C	$3\infty[C^t]^{-Hc}$	Cub. $Fd\bar{3}m$	a=3.56679Å (20°C) Z=8	C(8a)	$3\infty[C^t]^{-Hc}$ DIAMOND	Wyckoff,1,25-26;Pov.,192;LF, 253.
ELECTRUM	(Au,Ag)	$[Au,Ag]^c$	Cub. $Fm\bar{3}m$	a=4.08Å Z=4	(Au,Ag)(4a)	$[Cu]^c$ COPPER	Min. Abs.,88M/4285;Pov.,193.
FERCHROMIDE	$Cr_{1.5}Fe_{0.2}$	$[Cr_{1.5},Fe_{0.2},\square_{0.3}]^b$	Cub. $Pm\bar{3}m$	a=2.859Å Z=1			ZVMO,1986,115,355-360;Am. Min.,1988,73,191(Abs.);Hölzel,3.
GOLD	Au	$[Au]^c$	Cub. $Fm\bar{3}m$	a=4.07825Å (25°C) Z=4	Au(4a)	$[Cu]^c$ COPPER	Wyckoff,10;Pov.,739,193;J.Appl. Cryst.,1968,1,123-124.
GRAPHITE-2H	C	$2\infty[C^{tr}]^{Hh}$	Hex. $P6_3/mmc$	a=2.456Å c=6.709Å Z=4	C_I(2a) C_{II}(2b)	$2\infty[C^{tr}]^{Hh}$ GRAPHITE-2H	O'Keeffe & Hyde,436;Wyckoff, 1,26-27;Pov.,197;Hölzel,8; K/S,269;LF,223.
GRAPHITE-3R	C	$2\infty[C^{tr}]^{Hc}$	Trig. $R\bar{3}m$	a=2.456Å a_R=3.635Å c=10.044Å α=39°30′ Z=6 Z_R=2	C(6c)	$2\infty[C^{tr}]^{Hc}$ GRAPHITE-3R	Bokii,247;Pov.297;Wyckoff,1,28; Pov.,127.

Table 2

A(cont.)

NAME	CHEMICAL FORMULA	STRUCTURAL FORMULA	SPACE GROUP	UNIT CELL DIMENSIONS	EQUIVALENT POSITIONS	STRUCTURE TYPE	REFERENCES
INDIUM	In	$[In]^c$	Tet. I 4/mmm	a=3.244Å c=4.938Å Z=2	In(2a)	Dist.deriv. $[Cu]^c$ COPPER	Wyckoff,1,18;Am.Min.1967, 52,299(Abs.);Bokii,235-236.
IRIDIUM	Ir	$[Ir]^c$	Cub. $Fm\bar{3}m$	a=3.8383Å (30°C) Z=4	Ir(4a)	$[Cu]^c$ COPPER	Acta Cryst.,1968,24A,469-471;Pov.,192-193.
IRIDOSMINE	(Os,Ir)	$[Os,Ir]^h$	Hex. $P6_3/mmc$	a=2.63Å c=4.24Å Z=2	(Os,Ir)(2c)	$[Os]^h$ OSMIUM	Pov.,192;Wyckoff,1,11;LF,84;Str.Tab.,98.
IRON	Fe α-form	$[Fe]^b$	Cub. $Im\bar{3}m$	a=2.8665Å (25°C) Z=2	Fe(2a)	$[Fe]^b$ IRON	Wyckoff,1,15-16;Pov.,193;LF,135;Str.Tab.,96.
KAMACITE	(Fe,Ni) α-form	$[Fe,Ni]^b$	Cub. $Im\bar{3}m$	a=2.859Å Z=2			Am.Min.,1966,51,37-55;Hölzel,3,252.
LEAD	Pb	$[Pb]^c$	Cub. $Fm\bar{3}m$	a=4.9505Å Z=4	Pb(4a)	$[Cu]^c$ COPPER	Wyckoff,1,10;Pov.,193;Str.Tab.,95.
LEADAMALGAM	$Pb_{0.7}Hg_{0.3}$	$[Pb_{0.7},Hg_{0.3}]^c$	Tet. I 4/mmm	a=3.545Å Z= ? c=4.525Å			Am.Min.,1985,70,216(Abs.);Hölzel,2.
LONSDALEITE	C	$3\infty[C]^{t_1-Hh}$	Hex. $P6_3/mmc$	a=2.51Å c=4.12Å Z=4	C(4f)	$3\infty[C]^{t_1-Hh}$ LONSDALEITE	Am.Min.,1985,62,1579(Abs.);Pov.,192;Bokii,379-380;LF,253;Krebs,162;O´Keeffe & Hyde,300;Hölzel,8,253.
MERCURY	Hg	$[Hg]^c$ (T<-39°C)	Trig. $R\bar{3}m$	a=3.464Å a_R=2.9925Å c=6.677Å α=70°44'.6 Z=3 Z_R=1	Hg(3a)	Dist. deriv. $[Cu]^c$ COPPER	Wyckoff,1,17;Kostov,95;Bokii,236;Pearson(1967)48.
NICKEL	Ni	$[Ni]^c$	Cub. $Fm\bar{3}m$	a=3.5230Å Z=4	Ni(4a)	$[Cu]^c$ COPPER	Min.Mag.,1975,40,247-257;Pov.,193;Wyckoff,1,7-10.
OSMIRIDIUM	(Ir,Os)	$[Ir,Os]^c$	Cub. $Fm\bar{3}m$	a=3.88Å Z=4			Can.Min.,1973,12,104-112;Str.Tab.,98;Hölzel,6.
OSMIUM	Os	$[Os]^h$	Hex. $P6_3/mmc$	a=2.7352Å (20°C) c=4.3190Å Z=2	Os(2c)	$[Os]^h$ OSMIUM	Wyckoff,1,11;Str.Tab.,98.
PALLADIUM	Pd	$[Pd]^c$	Cub. $Fm\bar{3}m$	a=3.8898Å (25°) Z=4	Pd(4a)	$[Cu]^c$ COPPER	Wyckoff,1,10.
PARASCHACHNE-RITE	$Ag_{1.2}Hg_{0.8}$	$[Ag_{1.2}Hg_{0.8}]^h$	Orth. Cmcm ...	a=2.96Å Z=2 b=5.13Å c=4.83Å			Am.Min.,1973,58,347(Abs.);Bokii,301-302.
PLATINIRIDIUM	(Ir,Pt)	$[Ir,Pt]^c$	Cub. $Fm\bar{3}m$	a=3.8312Å Z=4	(Ir,Pt)(4a)	$[Cu]^c$ COPPER	RRW,483;Pov.,192-193.
PLATINUM	Pt	$[Pt]^c$	Cub. $Fm\bar{3}m$	a=3.9231Å (25°) Z=4	Pt(4a)	$[Cu]^c$ COPPER	Wyckoff,1,10.
RHODIUM	Rh	$[Rh]^c$	Cub. $Fm\bar{3}m$	a=3.8031Å (25°) Z=4	Rh(4a)	$[Cu]^c$ COPPER	Wyckoff,1,10.

Table 3

A(cont.)

NAME	CHEMICAL FORMULA	STRUCTURAL FORMULA	SPACE GROUP	UNIT CELL DIMENSIONS	EQUIVALENT POSITIONS	STRUCTURE TYPE	REFERENCES
ROSICKÝITE	S γ-form	$\{g\}[S_8]^{-h}$	Mon. P2/c	a=8.442Å β=124°.98 b=13.025Å Z=4(mol.) c=9.356Å	S_{I-VIII}(4g)	$\{g\}[S_8]^{-h}$ ROSICKÝITE	Acta Cryst.,1974,B30,1396-1401.
RUTHENIRIDOSMINE	(Os,Ir,Ru)	$[Os,Ir,Ru]^h$	Hex. $P6_3/mmc$	a=2.71Å c=4.30Å Z=2	(Os,Ir,Ru)(2c)	$[Os]^h$ OSMIUM	Str.Tab.,98;Pov.,192;LF,84; Hölzel,6.
RUTHENIUM	Ru	$[Ru]^h$	Hex. $P6_3/mmc$	a=2.7038Å c=4.2816Å Z=2	Ru(2c)	$[Os]^h$ OSMIUM	Wyckoff,1,11;Str.Tab.,98; Am.Min.,1976,61,177(Abs.).
RUTHENOSMIRIDIUM	(Ir,Os,Ru)	$[Ir,Os,Ru]^c$	Cub. $Fm\bar{3}m$	a=3.83Å Z=4	(Ir,Os,Ru)(4a)	$[Cu]^c$ COPPER	CIM SV 23,1981,83;Str.Tab., 98;Str.Tab.,6.
SCHACHNERITE	$Ag_{1.1}Hg_{0.9}$ β-form	$[Ag_{1.1}Hg_{0.9}]^h$	Hex. $P6_3/mmc$	a=2.978Å c=4.842Å Z=2	$(Ag_{1.1}Hg_{0.9})$(2c)	$[Os]^h$ OSMIUM	Am.Min.,1973,58,347(Abs.); Bokii,301-302;RRW,542.
SELENIUM	Se γ-form	$1\infty[Se^{[2]}]$	Trig. $P3_121$...	a=4.36Å c=4.96Å Z=3	Se(3a)	$1\infty[Se^{[2]}]$ SELENIUM	Str.Tab.,103;Wyckoff,1,36; Pov.,196;LF,201.
SILVER-2H	Ag	$[Ag]^h$	Hex. $P6_3/mmc$	a=2.94Å c=4.80Å Z=2	Ag(2c)	$[Os]^h$ OSMIUM	Am.Min.,1980,65,1069(Abs.); Hölzel,1.
SILVER-3C	Ag	$[Ag]^c$	Cub. $Fm\bar{3}m$	a=4.0862Å Z=4	Ag(4a)	$[Cu]^c$ COPPER	Wyckoff,1,10;Str.Tab.,94.
SILVER-4H	Ag	$[Ag]^{ch}$	Hex.	a=2.94Å c=10.11Å Z=4			Am.Min.,1980,65,1069(Abs.); Acta Cryst.,1969,B25,477-488; Wyckoff,1,14.
SULPHUR	S α-form	$\{g\}[S_8]$	Orth. Fddd	a=10.4646Å Z=16 b=12.8660Å c=24.4860Å	S_{I-IV}(32h)	$\{g\}[S_8]$ SULPHUR	Acta Cryst.,1980,C43,2260-2262;LF,177.
TAENITE	(Ni,Fe) γ-form	$[Ni,Fe]^c$	Cub. $Fm\bar{3}m$	a=7.168Å Z=32			Am.Min.,1966,51,37-55;Hölzel, 3;RRW,600.
TELLURIUM	Te	$1\infty[Te^{[2]}]$	Trig. $P3_121$...	a=4.4572Å c=5.9290Å Z=3	Te(3a)	$1\infty[Se^{[2]}]$ SELENIUM	Acta Cryst.,1967,23,670-671; Wyckoff,1,36;LF,201.
TIN	Sn β-form	$3\infty[Sn^t]^{t-Hc}$	Tet. $I4_1/amd$	a=5.8197Å (25°C) c=3.17488Å Z=4	Sn(4a)	Dist.deriv. $3\infty[C^t]^{t-Hc}$ DIAMOND	Wyckoff,1,28-29;LF,254; Str.Tab.,95.
ZINC	Zn	$[Zn]^h$	Hex. $P6_3/mmc$	a=2.6648Å (25°C) c=4.9967Å Z=2	Zn(2c) c/a=1.86 c/a(ideal)=1.63	Dist.deriv. $[Os]^h$ OSMIUM	Wyckoff,1,11;Bokii,258.

Table 4

AB

NAME	CHEMICAL FORMULA	STRUCTURAL FORMULA	SPACE GROUP	UNIT CELL DIMENSIONS	EQUIVALENT POSITIONS	STRUCTURE TYPE	REFERENCES
ALABANDITE	MnS α-form	$Mn^{o}[S]^{c}$	Cub. $Fm\bar{3}m$	a=5.214Å Z=4	S(4a) Mn(4b)	$Na^{o}[Cl]^{c}$ HALITE	RRW,9;Pov.,220-221;Wyckoff,1, 88;LF,85.
ALTAITE	PbTe	$Pb^{o}[Te]^{c}$	Cub. $Fm\bar{3}m$	a=6.452Å Z=4	Te(4a) Pb(4b)	$Na^{o}[Cl]^{c}$ HALITE	RRW,14;Pov.,211;Wyckoff,1,89; LF,85.
BETA-MOISSANITE	SiC	$C^{t}[Si]^{c}$	Cub. $F\bar{4}3m$	a=4.348Å Z=4	C(4a) Si(4c)	$Zn^{t}[S]^{c}$ SPHALERITE	Acta Cryst.,1969,B25,477-488; Wyckoff,1,113-114.
BRAGGITE	(Pt,Pd,Ni)S	$(Pt,Pd,Ni)^{sq}[S]^{N21\,s}$	Tet. $P4_2/m$	a=6.380Å c=6.570Å Z=8	Pd(2d) S(8K) $(Pt,Ni)_{I}$(2e) $(Pt,Ni)_{II}$(4j)	$(Pd,Ni)^{sq}[S]^{N21\,s}$ VYSOTSKITE	Acta Cryst.,1973,B29,1446-1451;Am.Min.,1978,63,832-839.
BREITHAUPTITE	NiSb	$Ni^{o}[Sb]^{n}$	Hex. $P6_3/mmc$	a=3.942Å c=5.155Å Z=2	Ni(2a) Sb(2c)	$Ni^{o}[As]^{n}$ NICKELINE	RRW,87;Pov.,203;LF,86.
BROMARGYRITE	AgBr	$Ag^{o}[Br]^{c}$	Cub. $Fm\bar{3}m$	a=5.77Å Z=4	Br(4a) Ag(4b)	$Na^{o}[Cl]^{c}$ HALITE	Str.Tab.,154-155;Pov.,730,637; LF,85.
BROMELLITE	BeO	$Be^{t}[O]^{n}$	Hex. $P6_3mc$	a=2.675Å c=4.341Å Z=2	Be(2b) O(2b)	$Zn^{t}[S]^{n}$ WURTZITE-2H	SR,53A,123;Pov.,272;LF,114.
BUNSENITE	NiO	$Ni^{o}[O]^{c}$	Cub. $Fm\bar{3}m$	a=4.1768Å Z=4	O(4a) Ni(4b)	$Na^{o}[Cl]^{c}$ HALITE	RRW,95;Pov.,271;LF,85.
CADMOSELITE	CdSe β-form	$Cd^{t}[Se]^{n}$	Hex. $P6_3mc$	a=4.299Å c=7.010Å Z=2	Cd(2b) Se(2b)	$Zn^{t}[S]^{n}$ WURTZITE-2H	Acta Cryst.,1977,A33,355-359; Am.Min.,1958,43,623(Abs.); LF,114.
CALOMEL	HgCl α-form	$1\infty[Hg^{[5]}Cl^{«(2Qs)f»}]$	Tet. I 4/mmm	a=4.4795Å c=10.9054Å Z=4	Hg(4e) Cl(4e)	$1\infty[Hg^{[5]}Cl^{«(2Qs)f»}]$ CALOMEL	Zeit.Krist.,1989,187,305-307; LF,203;Wyckoff,1,151-152.
CARLSBERGITE	CrN	$Cr^{o}[N]^{c}$	Cub. $Fm\bar{3}m$	a=4.16Å Z=4	N(4a) Cr(4b)	$Na^{o}[Cl]^{c}$ HALITE	Hölzel,5;Bokii,370;Wyckoff,1,86.
CAROBBIITE	KF	$[KF]^{Qs}$	Cub. $Fm\bar{3}m$	a=5.34Å Z=4	F(4a) K(4b)	$[KF]^{Qs}$ CAROBBIITE	RRW,109;Pov.,661-662; Wyckoff,1,86-87;LF,137.
CHEREPANOVITE	RhAs	$Rh^{o}[As]^{n}$	Orth. Pnma	a=5.70Å Z=4 b=3.59Å c=6.00Å			Am.Min.,1986,71,1544(Abs.); Bokii,377.
CHLORARGYRITE	AgCl	$Ag^{o}[Cl]^{c}$	Cub. $Fm\bar{3}m$	a=5.556Å Z=4	Cl(4a) Ag(4b)	$Na^{o}[Cl]^{c}$ HALITE	RRW,114;Pov.,637;Wyckoff,1, 86;LF,85.
CINNABAR	HgS	$1\infty[Hg^{o}S^{«Tv»}]$	Trig. $P3_121$	a=4.149Å Z=3 c=9.495Å	Hg(3a) S(3b)	$1\infty[Hg^{o}S^{«Tv»}]$ CINNABAR	Wyckoff,1,98-99;Bull.Min.,1973, 96,218-219;LF,202.
CLAUSTHALITE	PbSe	$Pb^{o}[Se]^{c}$	Cub. $Fm\bar{3}m$	a=6.147Å Z=4	Se(4a) Pb(4b)	$Na^{o}[Cl]^{c}$ HALITE	RRW,133;Pov.,220-221;LF,85.
COLORADOITE	HgTe	$Hg^{t}[Te]^{c}$	Cub. $F\bar{4}3m$	a=6.448Å Z=4	Hg(4a) Te(4c)	$Zn^{t}[S]^{c}$ SPHALERITE	RRW,142;Pov.,211;Am.Min.1949, 34,342-382;Wyckoff,1,108-110; LF,110.
COOPERITE	(Pt,Pd,Ni)S	$(Pt,Pd,Ni)^{sq}[S]^{Qs}$	Tet. $P4_2/mmc$	a=3.465Å c=6.104Å Z=2	(Pt,Pd,Ni)(2c) S(2e)	$Pt^{sq}[S]^{Qs}$ COOPERITE	Am.Min.,1978,63,832-839; Wyckoff,1,139;LF,142.

Table 5

AB(cont.)

NAME	CHEMICAL FORMULA	STRUCTURAL FORMULA	SPACE GROUP	UNIT CELL DIMENSIONS	EQUIVALENT POSITIONS	STRUCTURE TYPE	REFERENCES
COVELLITE	CuS	$Cu_2^{t}Cu^{tr}[S_3]^{(31n)s}$	Hex. $P6_3/mmc$	a=3.7938Å b=16.341Å Z=6	Cu^{tr}(2d) S_{II}(4e) Cu^{t}(4f) S_{I}(2c)	$Cu_2^{t}Cu^{tr}[S_3]^{(31n)s}$ COVELLITE	Am.Min.,1976,61,996-1000; Wycoff,1,145-146;LF,156.
CUPALITE	(Cu,Zn)Al		Orth.	a=6.95Å Z=10 b=4.16Å c=10.04Å			ZVMO,1985,114,90-100; Hölzel,3.
EMPRESSITE	AgTe		Orth. Pmnm...	a=8.90Å Z=16 b=20.07Å c=4.62Å			Am.Min.,1964,49,325-338; Pov.,211;Hölzel,14.
FREBOLDITE	CoSe γ-form	$Co^{o}[Se]^{n}$	Hex. $P6_3/mmc$	a=3.61Å c=5.28Å Z=2	Co(2a) Se(2c)	$Ni^{o}[As]^{n}$ NICKELINE	Str.Tab.,126;Pov.,220;LF,86.
GALENA	PbS	$Pb^{o}[S]^{c}$	Cub. $Fm\bar{3}m$	a=5.936Å Z=4	S(4a) Pb(4b)	$Na^{o}[Cl]^{c}$ HALITE	RRW,228;Pov.,220;Wyckoff,1, 89;LF,85.
GREENOCKITE	CdS β-form	$Cd^{t}[S]^{n}$	Hex. $P6_3mc$	a=4.142Å c=6.724Å Z=2		$Zn^{t}[S]^{n}$ WURTZITE-2H	RRW,247;Pov.,223;LF,114; Hölzel,22.
GRICEITE	LiF	$Li^{o}[F]^{c}$	Cub. $Fm\bar{3}m$	a=4.0293Å Z=4	F(4a) Li(4b)	$Na^{o}[Cl]^{c}$ HALITE	Can.Min.,1989.27,125-127; Wyckoff,1,85-94.
HALITE	NaCl	$Na^{o}[Cl]^{c}$	Cub. $Fm\bar{3}m$	a=5.62779Å (18°C) Z=4	Cl(4a) Na(4b)	$Na^{o}[Cl]^{c}$ HALITE	Wyckoff.1,85-94;Pov.,637-638; LF,85.
HAWLEYITE	CdS α-form	$Cd^{t}[S]^{c}$	Cub. $F\bar{4}3m$	a=5.818Å Z=4	Cd(4a) S(4c)	$Zn^{t}[S]^{c}$ SPHALERITE	RRW,262;Pov.,222;Am.Min., 1955,40,555-559;LF,110.
HERZENBERGITE	SnS	$Sn^{o}[S]^{c}$	Orth. Pnma	a=11.180Å Z=4 b=3.982Å c=4.329Å	Sn(4c) S(4c)	Dist.deriv. $Na^{o}[Cl]^{c}$ HALITE	Acta Cryst.,1981,B37,1903-1905; K/S,25;Pov.,261-262;Hölzel,23.
HONGSHIITE	PtCu	$[PtCu]^{c}$ (Subs.deriv. Copper)	Trig. R32...	a=10.713Å a_R=7.589Å c=13.192Å. α=89°47' Z=45 Z=15		$[PtCu]^{c}$ HONGSHIITE	Am.Min.,1984,69,411-412(Abs.) Bokii,299;Hölzel,7.
HYPERCINNABAR	HgS		Hex.	a=7.01Å Z=!2 c=14.13Å			Am.Min.,1978,63,1143-1152; Hölzel,22.
INGODITE	Bi(Te,S)		Hex. $P\bar{3}m1$...	a=4.248Å c=23.22Å Z=9 (for Bi_2TeS)			Am.Min.,1982,67,855(Abs.);Am. Min.,1985,70,220(Abs.);Hölzel,3 0
IODARGYRITE	AgI β-form	$Ag^{t}[I]^{n}$	Hex. $P6_3mc$	a=4.58Å Z=2 c=7.49Å		$Zn^{t}[S]^{n}$ WURTZITE-2H	Hölzel,48;Pov.637;Wyckoff,1, 111-112;LF,114.
KHAMRABAEVITE	(Ti,V,Fe)C	$(Ti,V,Fe)^{o}[C]^{c}$	Cub. $Fm\bar{3}m$	a=4.319Å Z=8			Hölzel.4;Bokii,357;ZVMO,1984, 113,69.
KLOCKMANNITE	CuSe	$Cu_2^{t}Cu^{tr}[Se_3]^{(31n)s}$	Hex. $P6_3/mmc$	a=3.939Å Z=6 c=17.25Å		$Cu_2^{t}Cu^{tr}[Se_3]^{(31n)s}$ COVELLITE	Min.Abs.,82-M 0177;Pov.,265-266;Am.Min.,1949,34,435-445.

Table 6

AB(cont.)

NAME	CHEMICAL FORMULA	STRUCTURAL FORMULA	SPACE GROUP	UNIT CELL DIMENSIONS	EQUIVALENT POSITIONS	STRUCTURE TYPE	REFERENCES
KOTULSKITE	Pd(Te,Bi)	Pd^{o}[Te,Bi]^{h}	Hex. $P6_3/mmc$	a=4.19Å c=5.67Å Z=2	Pd(2a) (Te,Bi)(2c)	Ni^{o}[As]^{h} NICKELINE	Am.Min.,1963,48,1181(Abs.); Pov.,210;Hölzel,36;LF,203.
KUZMINITE	Hg(Br,Cl)	$_{1\infty}$[Hg$^{[5]}$ (Br,Cl)$^{«(2Qs)fs}$]	Tet. I 4/mmm	a=4.597Å c=11.034Å Z=4	Hg(4e) (Br,Cl)(4e)	$_{1\infty}$[Hg$^{[5]}$Cl$^{«(2Qs)fs}$] CALOMEL	Am.Min.,1988,73,192(Abs.); Wyckoff,1,151-152.
LANGISITE	(Co,Ni)As	(Co,Ni)^{o}[As]^{h}	Hex. $P6_3/mmc$	a=3.538Å c=5.127Å Z=2	(Co,Ni)(2a) As(2c)	Ni^{o}[As]^{h} NICKELINE	Can.Min.,1969,9,597-616;RRW, 341;Pov.,203;Hölzel,25.
LIME	CaO	Ca^{o}[O]^{c}	Cub. $Fm\bar{3}m$	a=4.81Å Z=4	O(4a) Ca(4b)	Na^{o}[Cl]^{c} HALITE	Pov.,271;RRW,357;LF,85.
LITHARGE	PbO γ-form	O^{t}[Pb]^{c}	Tet. P4/nmm	a=3.975Å Z=2 c=5.023Å	O(2a) Pb(2c)	O^{t}[Pb]^{c} LITHARGE	Wyckoff,1,134-135;Pov.,310; LF,117.
MÄKINENITE	NiSe γ-form	$_{3\infty}$[Ni$^{[5]}$,Se]	Trig. $R\bar{3}m$	a=10.01Å a_R=5.8834Å c=3.28Å α=116°31' Z=9 Z_R=3	Ni(9b) Se(9b)	$_{3\infty}$[Ni$^{[5]}$S] MILLERITE	Wyckoff,1,122;Am.Min.,1965,50, 519-520(Abs.);Pov.,247-248.
MANGANOSITE	MnO	Mn^{o}[O]^{c}	Cub. $Fm\bar{3}m$	a=4.446Å Z=4	O(4a) Mn(4b)	Na^{o}[Cl]^{c} HALITE	Acta Cryst.,1980,A36,904-915; Pov.,271;LF,85.
MARSHITE	CuI	Cu^{t}[I]^{c}	Cub. $F\bar{4}3m$	a=6.047Å Z=4	Cu(4a) I(4c)	Zn^{t}[S]^{c} SPHALERITE	RRW,383-384;Pov.,637.
MASSICOT	PbO β-form	O^{t}[Pb]^{c}	Orth. Pbcm	a=5.8931Å Z=4 b=5.4904Å c=4.7528Å	Pb(4d) O(4d)	Dist.deriv. O^{t}[Pb]^{c} LITHARGE	Acta Cryst.,1985,C41,1281-1284; Pov.,310;LF,117.
MATRAITE	ZnS	Zn^{t}[S]^{c}	Trig. R3m	a=3.8Å c=9.4Å Z=3		Dist.deriv. Zn^{t}[S]^{c} SPHALERITE	Am.Min.,1960,45,1131(Abs.); RRW,386;Am.Min.,1955,40,192-200.
METACINNABAR	HgS	Hg^{t}[S]^{c}	Cub. $F\bar{4}3m$	a=5.8717Å Z=4	Hg(4a) S(4c)	Zn^{t}[S]^{c} SPHALERITE	SR,29,67-68;Pov.,222;RRW,36; LF,110.
MIERSITE	(Ag,Cu)I α-form	(Ag,Cu)^{t}[I]^{c}	Cub. $F\bar{4}3m$	a=6.491Å Z=4	(Ag,Cu)(4a) I(4c)	Zn^{t}[S]^{c} SPHALERITE	RRW,405-406;Pov.,637;LF,110
MILLERITE	NiS β-form	$_{3\infty}$[Ni$^{[5]}$S]	Trig. R3m	a=9.612Å a_R=5.655Å c=3.259Å α=116°36' Z=9 Z_R=3	Ni(9b) S(9b)	$_{3\infty}$[Ni$^{[5]}$S] MILLERITE	Wyckoff,1,122;Pov.,247-248; LF,273.
MODDERITE	(Co,Fe)As	(Co,Fe)^{o}[As]^{h}	Orth. Pnma	a=5.416Å Z=4 b=3.379Å c=6.011Å	(Co,Fe)(4c) As(4c)	Dist.deriv. Ni^{o}[As]^{h} NICKELINE	SR,41A,19-20;Pov.,203;Bokii, 377;LF,203.
MOISSANITE-5H	SiC	C^{t}[Si]^{ccchh}	Hex.	a=3.03Å c=12.37Å Z=6			Am.Min.,1976,61,1054(Abs.); Hölzel,8;Acta Cryst.,1969,B25, 477-488.
MOISSANITE-6H	SiC α-form	C^{t}[Si]^{cch}	Hex. $P6_3mc$	a=3.075Å Z=6 c=15.06Å	Si_I(2a) C_I(2a) Si_{II-III}(2b) C_{II-III}(2b)		Acta Cryst.,1969,B25,477488; RRW,411;Hölzel,8:SR,13,166.
MONTEPONITE	CdO	Cd^{o}[O]^{c}	Cub. $Fm\bar{3}m$	a=4.70Å Z=4	O(4a) Cd(4b)	Na^{o}[Cl]^{h} HALITE	Pov.,271-272;Hölzel,58.

Table 7

AB(cont.)

NAME	CHEMICAL FORMULA	STRUCTURAL FORMULA	SPACE GROUP	UNIT CELL DIMENSIONS	EQUIVALENT POSITIONS	STRUCTURE TYPE	REFERENCES
MONTROYDITE	HgO	$_{1\infty}[Hg^{[2]}O]$	Orth. Pnma	a=6.612Å Z=4 b=5.520Å c=3.521Å	Hg(4c) O(4c)	$_{1\infty}[Hg^{[2]}O]$ MONTROYDITE	SR,20,266-268;Pov.,300-301;Hölzel,59.
NANTOKITE	CuCl	$Cu^{t}[Cl]^{c}$	Cub. F$\bar{4}$3m	a=5.418Å Z=4	Cu(4a) Cl(4c)	$Zn^{t}[S]^{c}$ SPHALERITE	RRW,426;Pov.,637;LF,110.
NEVSKITE	Bi(Se,S)	$Bi^{o}[Se,S]^{c}$	Trig. P$\bar{3}$m1...	a=4.197Å Z=6 c=22.80Å			Am.Min.,1985,70,875(Abs.); Wyckoff,1,86.
NICKELINE	NiAs	$Ni^{o}[As]^{h}$	Hex. $P6_3/mmc$	a=3.602Å Z=2 c=5.009Å	Ni(2a) As(2c)	$Ni^{o}[As]^{h}$ NICKELINE	Wyckoff, 1,122-124;Pov., 203;RRW,434;LF,86.
NIGGLIITE	PtSn	$Pt^{o}[Sn]^{h}$	Hex. $P6_3/mmc$	a=4.100Å Z=2 c=5.432Å	Pt(2c) Sn(2c)	$Ni^{o}[As]^{h}$ NICKELINE	Min.Mag.,1972,38,794-800; RRW,436;Pov.,195;LF,86.
NININGERITE	(Mg,Fe,Mn)S	$(Mg,Fe,Mn)^{o}[S]^{c}$	Cub. Fm$\bar{3}$m	a=5.17Å Z=4	S(4a) (Mg,Fe,Mn)(4b)	$Na^{o}[Cl]^{c}$ HALITE	Am.Min.,1967,52,925(Abs.); Str.Tab.,123;LF,85.
OLDHAMITE	(Ca,Mg)S	$(Ca,Mg)^{o}[S]^{c}$	Cub. Fm$\bar{3}$m	a=5.686Å Z=4	S(4a) (Ca,Mg)(4b)	$Na^{o}[Cl]^{c}$ HALITE	RRW,446;Pov.,220; Hölzel,23.
OSBORNITE	TiN	$Ti^{o}[N]^{c}$	Cub. Fm$\bar{3}$m	a=4.235Å Z=4	N(4a) Ti(4b)	$Na^{o}[Cl]^{c}$ HALITE	RRW,452;Pov.,200; Min.Mag.,1941,26,36-44.
PARAREALGAR	AsS	$\{g\}[As_4{}^{[3n]}S_4]$	Mon. $P2_1/c$	a=9.909Å β=97°29' b=9.655Å Z=16? c=8.502Å	$As_{1\text{-}iv}(4e)$ $S_{1\text{-}iv}(4e)$	$\{g\}[As_4{}^{[3n]}S_4]$ PARAREALGAR	Am.Min.,1995,,80,400-403; Can.Min.,1980,18,525-527; Hölzel,48.
PERICLASE	MgO	$Mg^{o}[O]^{c}$	Cub. Fm$\bar{3}$m	a=4.2121Å Z=4	O(4a) Mg(4b)	$Na^{o}[Cl]^{c}$ HALITE	Am.Min.,1976,61,266-271; Pov.,271;Wyckoff,1,88;LF,85.
POLARITE	Pd(Pb,Bi)	$[Pd(Pb,Bi)]^{c}$	Orth. $Ccm2_1$?	a=7.191Å Z=16 b=8.693Å c=10.681Å			Am.Min.,1970,55,1810(Abs.). Hölzel,suppl.;Pov.,194-195; ZVMO,1969,98,708.
POLHEMUSITE	(Zn,Hg)S	$(Zn,Hg)^{t}[S]^{c}$	Tet. P4/n...	a=8.71Å c=14.74Å Z=24...		Disordered deriv. $Zn^{t}[S]^{c}$ SPHALERITE	Am.Min.,1978,63,1153-1161; Hölzel,22;K/S,42.
POTARITE	PdHg	$[PdHg]^{c}$	Tet. P4/mmm	a=3.020Å Z=1 c=3.706Å	Hg(1a) Pd(1d)	$[AuCu]^{c}$ TETRAAURICUPRIDE	Am.Min.,1960,45,1093-1097; Pov.,194-195.
PYRRHOTITE-7C	$Fe_{1\text{-}x}S$ β-form	$(Fe_{1\text{-}x},\square_x)^{o}[S]^{h}$	Hex. $P6_3/mmc$	a=3.44Å c=5.69Å Z=2	$Fe_{1\text{-}x}(2a)$ S(2c)	Defect deriv. $Ni^{o}[As]^{h}$ NICKELINE	K/S 27-28;Str.Tab.,120; Pov.,225.
REALGAR	AsS β-form	$\{g\}[As_4{}^{[3n]}S_4]$	Mon. $P2_1/n$	a=9.325Å β=106°23' b=13.571Å Z=16 c=6.587Å	$As_{1\text{-}iv}(4e)$ $S_{1\text{-}iv}(4e)$	$\{g\}[As_4{}^{[3n]}S_4]$ REALGAR	Zeit.Krist.,1972,136,48-65; Wyckoff,1,174-177;Pov.,242.
REALGAR (High)	AsS α-form		Mon.?				Am.Min.,1970,55,1338-1344.
ROMARCHITE	SnO		Tet. P4/nmm	a=3.79Å c=4.83Å Z=2			Can.Min.,1971,10,916(Abs.); Hölzel,58;Bokii,382-383.

Table 8

AB(cont.)

NAME	CHEMICAL FORMULA	STRUCTURAL FORMULA	SPACE GROUP	UNIT CELL DIMENSIONS	EQUIVALENT POSITIONS	STRUCTURE TYPE	REFERENCES
RUTHENARSENITE	(Ru,Ni)As	$(Ru,Ni)^{o}[As]^{h}$	Orth. Pnma	a=5.628Å Z=4 b=3.239Å c=6.184Å		Dist.deriv. $Ni^{o}[As]^{h}$ NICKELINE	Can.Min.,1971,12,280-284; Hölzel,25;Bokii,377;K/S,158.
SAKURAIITE	(Cu,Zn, Fe,In,Sn)S	$Cu_2{}^{t}(Zn,Fe)^{t}$ $(In,Sn)^{t}[S_4]^{c}$	Cub. Pm3m	a=5.4563Å Z=4		$Cu_2{}^{t}Fe^{t}Sn^{t}[S_4]^{c}$ STANNITE	Can.Min.,1986,24,679-683;Hölzel, 20;Str.Tab.,115116;Pov.,228; Garcia Orta,Sér.Geol.,1986,9,41- 53;Min.Abs.,88M/1824.
SAL AMMONIAC	$(NH_4)Cl$ α-form	$(NH_4)^{cb}[Cl]^{Cs}$	Cub. Pm3m	a=3.8758Å Z=1	$(NH_4)(1a)$ Cl(1b)	$(NH_4)^{cb}[Q]^{Cs}$ SAL AMMONIAC	RRW,533;Pov.,638; Wyckoff,1,103.
SEDERHOLMITE	NiSe β-form	$Ni^{o}[Se]^{h}$	Hex. $P6_3/mmc$	a=3.65Å Z=2 c=5.34Å	Ni(2a) Se(2c)	$Ni^{o}[As]^{h}$ NICKELINE	Am.Min.,1965,50,519-520(Abs.); Pov.,225;Str.Tab.,120.
SOBOLEVSKITE	PdBi		Hex. $P6_3/mmc$	a=4.23Å Z=2 c=5.69Å			Am.Min.,1976,61,1054(Abs.); Hölzel,25.
SPHALERITE	ZnS α-form	$Zn^{t}[S]^{c}$	Cub. $F\bar{4}3m$	a=5.4093Å (26°) Z=4	Zn(4a) S(4c)	$Zn^{t}[S]^{c}$ SPHALERITE	Wyckoff,1,108-110;Pov.,222.
STIBARSEN	SbAs	$2\infty[(Sb,As)^{[3n]}]^{-Hc}$	Trig. $R\bar{3}m$	a=4.02Å a_R=4.29Å c=10.81Å α=55°53' Z=6 Z_R=2		$2\infty[As^{[3n]}]^{-Hc}$ ARSENIC	Pov.,197-198;Hölzel,8;Wyckoff,1, 321;Str.Tab.,99;Am.Min.,1974,59, 133,(Abs.).
STILLEITE	ZnSe	$Zn^{t}[Se]^{c}$	Cub. $F\bar{4}3m$	a=5.67Å Z=4	Zn(4c) Se(4c)	$Zn^{t}[S]^{c}$ SPHALERITE	RRW,583:Pov.,222;Acta Cryst., 1980,A36,482-490.
STISTAITE	SnSb	$Sn^{t}[Sb]^{c}$	Trig.	a=8.648Å Z=? c=10.675Å			Min.Abs.,81-4376;Pov.,759;Am. Min.,1971,56,358(Abs.);Hölzel,23.
STUMPFLITE	Pt(Sb,Bi)	$Pt^{o}[Sb,Bi]^{h}$	Hex. $P6_3/mmc$	a=4.175Å Z=5 c=5.604Å	Pt(2a) (Sb,Bi)(2c)	$Ni^{o}[As]^{h}$ NICKELINE	Am.Min.,1974,59,21(Abs.)
SUDBURYITE	(Pd,Ni)Sb	$(Pd,Ni)^{o}[Sb]^{h}$	Hex. $P6_3/mmc$	a=4.083Å Z=2 c=5.502Å	(Pd,Ni)(2a) Sb(2c)	$Ni^{o}[As]^{h}$ NICKELINE	Min.Abs.,80-1317;Bokii,376.
SYLVITE	KCl	$K^{o}[Cl]^{c}$	Cub. $Fm\bar{3}m$	a=6.2910Å Z=4	Cl(4a) K(4b)	$Na^{o}[Cl]^{c}$ HALITE	RRW,595;Pov.,637-638;Str.Tab., 154.
TANTALCARBIDE	TaC	$C^{o}[Ta]^{c}$	Cub. $Fm\bar{3}m$	a=4.4570Å Z=4	Ta(4a) C(4b)	$Na^{o}[Cl]^{c}$ HALITE	SR,26,104;Am.Min.,1962,47, 786-787; Str.Tab.,97.
TENORITE	CuO	$Cu^{sq}[O]^{Cs}$	Mon. C2/c	a=4.6837Å β=99.54° b=3.4226Å Z=4 c=5.1288Å	Cu(4c) O(4e)	Dist.deriv. $Pt^{sq}[S]^{Cs}$ COOPERITE	Acta Cryst.,1970,B26,8-15; Pov.,272;LF,143.
TETRAAURICUPRIDE	CuAu	$[CuAu]^{c}$ (Subs.deriv. Copper)	Tet. C4/mmm (or P4/mmm)	a=3.98Å c=3.72Å Z=2	Cu(2a) Au(2d)	$[AuCu]^{c}$ TETRAAURI- CUPRIDE	Am.Min.,1983,68,1250(Abs.); Pov.,194;SB,1,484;Bokii,294-296.

Table 9

AB(cont.)

NAME	CHEMICAL FORMULA	STRUCTURAL FORMULA	SPACE GROUP	UNIT CELL DIMENSIONS	EQUIVALENT POSITIONS	STRUCTURE TYPE	REFERENCES
TETRAFERROPLATI-NUM	PtFe	$[PtFe]^{c}$ (Subs.deriv. Copper)	Tetr. C4/mmm	a=3.850Å c=3.693Å Z=2	Pt(2a) Fe(2d)	$[CuAu]^{c}$ TETRAAURICU-PRIDE	Hölzel,6;Pov.,193;Can.Min., 1975,13,117-126.
TETRATAENITE	FeNi	$[FeNi]^{c}$ (Subs.deriv. Copper)	Tetr. P4/mmm	a=2.533Å c=3.582Å Z=1	Fe(1a) Ni(1d)	$[CuAu]^{c}$ TETRAAURICU-PRIDE	Am.Min.,1980,65,624-630; Bokii,294-296.
TIEMANNITE	HgSe	$Hg^{t}[Se]^{c}$	Cub. $F\bar{4}3m$	a=6.073Å Z=4	Hg(4a) Se(4c)	$Zn^{t}[S]^{c}$ SPHALERITE	SR,20,152;Str.Tab.,113-114; LF,110.
TROILITE	FeS	$Fe^{o}[S]^{h}$	Hex. P 62c	a=5.962Å c=11.750Å Z=12	Fe(12i) S_{II}(4f) S_{I}(2a) S_{III}(6h)	Dist.deriv. $Ni^{o}[As]^{h}$ NICKELINE	SR,20,129-130;Acta Cryst., 1982,B38,1877-1887.
TSUMOITE	BiTe	$[BiTe]^{c}$	Trig. $P\bar{3}m1$	a=4.422Å Z=6 c=24.05Å	Te_{I-II}(2d) Bi_{I}(2c) Te_{III}(2c) Bi_{II-III}(2d)		Am.Min.,1978,63,1162-1165.
VILLIAUMITE	NaF	$Na^{o}[F]^{c}$	Cub. $Fm\bar{3}m$	a=4.633Å Z=4	F(4a) Na(4b)	$Na^{o}[Cl]^{c}$ HALITE	RRW,652-653;Pov.,661-662. Am.Min.,1970,55,126-134.
VULCANITE	CuTe	$Cu^{t}[Te]^{c}$	Orth. Pmnm	a=4.09Å Z=2 b=6.95Å c=3.15Å		Dist.deriv. $O^{t}[Pb]^{c}$ LITHARGE	Am.Min.,1961,46,258-268; K/S,25-27;Pov.,215;Str.Tab., 127.
VYSOTSKITE	(Pd,Ni)S	$(Pd,Ni)^{sq}[S]^{N21s}$	Tet. $P4_2/m$	a=6.429Å c=6.611Å Z=8	$(Pd,Ni)_{I}$(2e) S(8k) $(Pd,Ni)_{II}$(2c) $(Pd,Ni)_{III}$(4j)	$(Pd,Ni)^{sq}[S]^{N21s}$ VYSOTSKITE	Acta Cryst.,1985,C41,1829-1830;Am.Min.,1978,63,832-839;LF,147;Wyckoff,1,142-143.
WAIRAUITE	CoFe	$[CoFe]^{b}$ (Subs.deriv.Iron)	Cub. $Pm\bar{3}m$	a=2.86Å Z=1	Co(1a) Fe(1b)	$[FeCo]^{b}$ WAIRAUITE	Pov.,193,117;Min.Mag.,1964,33, 942-948;LF,136;Wyckoff,1,15.
WESTERVELDITE	(Fe,Ni,Co)As	$(Fe,Ni,Co)^{o}$ $[As]^{h}$	Orth. Pmcn	a=3.46Å Z=4 b=5.97Å c=5.33Å	(Fe,Ni,Co)(4c) As(4c)	Dist.deriv. $Ni^{o}[As]^{h}$ NICKELINE	Am.Min.,1972,57,354-363; Wyckoff,1,126-128;Bokii,377; RRW,668.
WURTZITE-2H	ZnS β-form	$Zn^{t}[S]^{h}$	Hex. $P6_3mc$	a=3.811Å c=6.234Å Z=2	Zn(2b) S(2b)	$Zn^{t}[S]^{h}$ WURTZITE-2H	Wyckoff,1,111-112:RRW,676-677;Pov.,223;Bokii,379-380; Pearson,1962,381;LF,114.
WÜSTITE	FeO	$Fe^{o}[O]^{c}$	Cub. $Fm\bar{3}m$	a=4.31Å Z=4	O(4a) Fe(4b)	$Na^{o}[Cl]^{c}$ HALITE	Pov.,271;Acta Cryst.,1982,B38, 1451-1456;Pov.,271;LF,85; Wyckoff,1,86-87.
ZHANGHENGITE	CuZn	$Cu^{cb}[Zn]^{cs}$	Cub. Pm3m	a=2.945Å Z=1	Cu(1a) Zn(1b)	$(NH_4)^{cb}[Cl]^{c}$ SAL AMMONIAC	Wyckoff,1,104-105;Acta Min. Sin.,1986,6,220-223.
ZINCITE	(Zn,Mn)O	$(Zn,Mn)^{t}[O]^{h}$	Hex. $P6_3mc$	a=3.242Å Z=2 c=5.176Å	(Zn,Mn)(2b) O(2b)	$Zn^{t}[S]^{h}$ WURTZITE-2H	RRW,687;Wyckoff,1,111-112; Pov.,272.

Table 10

AB_2

NAME	CHEMICAL FORMULA	STRUCTURAL FORMULA	SPACE GROUP	UNIT CELL DIMENSIONS	EQUIVALENT POSITIONS	STRUCTURE TYPE	REFERENCES
ACANTHITE	Ag_2S	$3\infty[Ag_2{}^{[2/3]}S^{t}]$	Mon. $P2_1/n$	a=4.23Å β=99°35' b=6.91Å Z=4 c=7.87Å	$Ag_{I-II}(4e)$ S(4e)	Dist.deriv. $3\infty[Cu_2{}^{[2]}O^{t}]$ CUPRITE	Can.Min.,1974,12,365-369; Pov.,241;Wyckoff,1,334-335; K/S,49;SR,32A,123.
AKHTENSKITE	MnO_2 ε-form		Hex. $P6_3/mmc$	a=2.85Å Z=1 c=4.48Å			Am.Min.,1983,68,473(Abs.); Hölzel,suppl..
AMAKINITE	$(Fe,Mg)(OH)_2$		Trig. $R\bar{3}m$	a=6.917Å a_R=6.276Å c=14.52Å α=66°52' Z=12 Z_R=4			Am.Min.,1962,47,1218(Abs.); Pov.,316;Str.Tab.,215;RRW,16.
ANATASE	TiO_2	$Ti^{o}[O_2]^{c}$	Tet. $I\,4_1/amd$	a=3.785Å Z=4 c=9.514Å	Ti(4a) O(8e)	$Ti^{o}[O_2]^{c}$ ANATASE	Wyckoff,1,253-254;Zeit.Krist., 1972,136,273-281;LF,90.
ANDUOITE	$(Ru,Os)As_2$	$(Ru,Os)^{o}$ $\{g\}[As_2{}^{«h»}]^{c}$	Orth. Pnnm	a=5.41Å Z=2 b=6.206Å c=3.01Å		$Fe^{o}\{g\}[S_2{}^{«h»}]^{c}$ MARCASITE	CIM SV23,1981,83;Am.Min., 1980,65,808-809(Abs.);Wyckoff, 1,355-356.
ANYUIITE	$AuPb_2$	$Au^{acb}[Pb_2]^{N2 1 T}$	Tet. $I\,4/mcm$	a=7.39Å Z=4 c=5.61Å	Au(4a) Pb(8h)	$Au^{acb}[Pb_2]^{N2 1 T}$ ANYUIITE	Min.Zhum.,1989,11(4),88; Bokii,330-331;SB,1,491.
ARGUTITE	GeO_2		Tet.	a=4.395Å Z=2 c=2.852Å			Hölzel,72;Am.Min.,1984,69,406 (Abs.).
ASHANITE	$(Nb,Ta,Fe,Mn,V)_4O_8$		Orth. Pcan	a=5.869Å Z=1 b=4.878Å c=5.216Å			Hölzel,75;Am.Min.,1981.66, 217(Abs.).
ASHOVERITE	$Zn(OH)_2$		Tet. $I\,4_1/amd$...	a=6.825Å Z=32 c=33.36Å			Min.Mag.,1988,52,699-702; Min.Abs.,89M/0932.
AUROSTIBITE	$AuSb_2$	$Au^{o}\{g\}[Sb_2]^{c}$	Cub. $Pa\bar{3}$	a=6.66Å Z=4	Au(4a) Sb(8c)	$Fe^{o}\{g\}[S_2]^{c}$ PYRITE	Pov.,206;Am.Min.,1952,37,461-469;Hölzel,32;Wyckoff,1,348;LF,171.
BADDELEYITE	ZrO_2	$Zr^{[7]}[O_2]^{Cs}$	Mon. $P2_1/c$	a=5.169Å β=99°15' b=5.232Å Z=4 c=5.341Å	Zr(4e) $O_{I-II}(4e)$	Dist.deriv. $Ca^{cb}[F_2]^{Cs}$ FLUORITE	Acta Cryst.,1965,18,983 991;SB,4, 9;Pov.,270;Wyckoff,1,244-245;SR,23, 342;Str.Tab.,24.
BAMBOLLAITE	$Cu(Se,Te)_2$	$Cu^{t}[(Se,Te)_2]^{c}$	Tet. $P4_2/n$	a=5.466Å Z=2 c=5.632Å			Can.Min.,1972,738-742;Am.Min.,1973, 58,805;Hölzel,32;K/S,168;Pov.,728.
BARRINGERI-TE	$(Fe,Ni)_2P$		Hex. $P\bar{6}2m$	a=5.868Å Z=3 c=3.458Å	$P_I(1b)$ $Fe_I(3f)$ $P_{II}(2c)$ $Fe_{II}(3g)$		SR,45A,86;Hölzel,5.
BEHOITE	$Be(OH)_2$ β-form		Orth. $P2_12_12_1$	a=4.64Å Z=4 b=7.05Å c=4.55Å			Am.Min.,1970,55,1-9; Pov.,316.
BELLIDOITE	Cu_2Se	$Cu_2{}^{t}[Se]^{c}$	Tet. $P4_2/n$	a=11.52Å Z=32 c=11.74Å			Econ.Geol.,1975,70,384-387; Hölzel,13;Bokii,385.
BERNDTITE-2T	SnS_2	$Sn^{o}[S_2]^{h}$	Trig. $P\bar{3}m1$	a=3.638Å Z=1 c=5.88Å		$Ni^{o}[Te_2]^{h}$ MELONITE	Am.Min.,1978,63,289-292; Hölzel,35;K/S,22;LF,98.
BERNDTITE-4H	SnS_2	$Sn^{o}[S_2]^{c/h}$	Hex. $P6_3mc$	a=3.65Å Z=2 c=11.8Å			Am.Min.,1973,58,347-349; Min.Mag.,1990,54,137;K/S,166.
BERZELIANITE	$Cu_{2-x}Se$	$(Cu_{1.8},\square_{0.2})^{t/tr}$ $[Se]^{c}$	Cub. $Fm\bar{3}m$	a=5.765Å Z=4	Cu(8c) Se(4af)	Defect. deriv. $Cu_2{}^{t}[S]^{c}$ DIGENITE	SR,42A,75;Pov.,232;Hölzel,13.

Table 11

AB_2(cont.)

NAME	CHEMICAL FORMULA	STRUCTURAL FORMULA	SPACE GROUP	UNIT CELL DIMENSIONS	EQUIVALENT POSITIONS	STRUCTURE TYPE	REFERENCES
BORISHANSKIITE	$Pd(As,Pb)_2$		Orth. $Ccm2_1$	a=7.18Å Z=10 ? b=8.62Å c=10.66			Am.Min.,1976,61,502-504; Hölzel,supl..
BROOKITE	TiO_2	$Ti^o[O_2]^{ch}$	Orth. Pbca	a=9.184Å Z=8 b=5.447Å c=5.145Å	Ti(8c) O_{I-II}(8c)	$Ti[O_2]^{ch}$ BROOKITE	Wyckoff,1,254-256;Pov.,309; LF,98.
BRUCITE	$Mg(OH)_2$	$Mg^o[(OH)_2]^n$	Trig. $P\bar{3}m1$	a=3.142Å Z=1 c=4.766Å	Mg(1a) H(2d) O(2d)	$Ni^o[Te_2]^n$ MELONITE	SR 32A,242;Pov.,324,55;LF,98.
CALAVERITE	$AuTe_2$	$Au^{[6]}[Te_2]^n$	Mon. C2/m	a=7.1947Å β=90.038° b=4.4146Å Z=2 c=5.0703Å	Au(2a) Te(4i)	Dist. deriv. $Ni^o[Te_2]^n$ MELONITE	Acta Cryst.,1988,B44,486-494; Wyckoff,1,337;Bokii,395;Pov.,210.
CARLINITE	Tl_2S	$S^o[Tl_2]^n$	Trig. R3	a=12.12Å c=18.175Å Z=27		Dist. deriv. $Ni^o[Te_2]^n$ MELONITE	Am.Min.,1975,60,559-565.
CASSITERITE	SnO_2	$Sn^o[O_2]^n$	Tet. $P4_2/mnm$	a=4.738Å Z=2 c=3.188Å	Sn(2a) O(4f)	$Ti^o[O_2]^n$ RUTILE	Wyckoff,1,250-251;Pov.,297;LF,88.
CATTIERITE	CoS_2	$Co^o\{g\}[S_2]^c$	Cub. $Pa\bar{3}$	a=5.5387Å Z=4	Co(4a) S(8c)	$Fe^o\{g\}[S_2]^c$ PYRITE	Zeit. Krist.,1979,150,163-167; Wyckoff,1,346-348;LF,171.
CERIANITE-(Ce)	CeO_2	$Ce^{cb}[O_2]^{Cs}$	Cub. $Fm\bar{3}m$	a=5.4110Å (26°C) Z=4	Ce(4a) O(8c)	$Ca^{cb}[F_2]^{Cs}$ FLUORITE	Pov.,269;Wyckoff,1,239-241;LF, 138.
CERVANTITE	Sb_2O_4	$Sb_2{}^o[O_4]^n$	Orth. $Pna2_1$	a=5.456Å Z=4 b=4.814Å c=11.787Å	Sb_{I-II}(4a) O_{I-IV}(4a)	$Pb^o[O_2]^n$ SCRUTINYITE	Acta Cryst.,1977,B33,1271-1273; Am.Min.,1962,47,1221(Abs.).
CHALCOCITE	Cu_2S	$Cu_2{}^{tr/t}[S]^n$	Mon. $P2_1/c$	a=15.246Å β=116.35° b=11.884Å Z=48 c=13.494Å	S_{I-XII}(4e) Cu_{I-XXIV}(4e)	$Cu_2{}^{tr/t}[S]^n$ CHALCOCITE	Zeit.Krist.,1979,150,299-320;K/S, 48;Hölzel,12;LF,304.
CHLOROMAGNESITE	$MgCl_2$	$Mg^o[Cl_2]^c$	Trig. $R\bar{3}m$	a=3.596Å a_R=6.22Å c=17.589Å α=33°36' Z=3 Z_R=1	Mg(3a) Cl(6c)	$Mg^o[Cl_2]^c$ CHLOROMA- GNESITE	Wyckoff,1,270-272;Pov.,644; LF,97.
CLINOBEHOITE	$Be(OH)_2$		Mon. $P2_1$	a=11.020Å β=98.90° b=4.746Å Z=12 c=8.646Å	Be_{I-VI}(2a) $(OH)_{I-XII}$(2a)		Zeit.Krist.,1988,185,612;Dokl. Akad.Sci.USSR,1989,305,95-98.
CLINOSAFFLORITE	$(Co,Fe,Ni)As_2$	$(Co,Fe,Ni)^o$ $\{g\}[As_2{}^{ch*}]^c$	Mon. $P2_1/n$	a=5.040Å β=90°13' b=5.862Å Z=2 c=3.139Å			Can.Min.,1971,10,877-881; Pov.,206.
COESITE	SiO_2	$3\infty[Si^tO_2{}^{[2]}]^c$	Mon. C2/c	a=7.135Å β=120.36° b=12.372Å Z=16 c=7.173Å	Si_{I-II}(8f) O_{II}(4e) O_I(4a) O_{III-V}(8f)	$3\infty[Si^tO_2{}^{[2]}]^c$ COESITE	Zeit.Krist.,1977,145,108-123;Am. Min.,1962,47,1292-1322;LF,265.
COTUNNITE	$PbCl_2$		Orth. Pbnm	a=9.030Å Z=4 b=7.608Å c=4.525Å	Pb(4c) Cl_{I-II}(4c)		Wyckoff,1,298-299;Pov.,636;Wells, 1962,338-339;Sov.Phys.Cryst., 1976,21,38-40;LF,157.
CRISTOBALITE	SiO_2	$3\infty[Si^tO_2{}^{[2]*c*}]^c$	Tet. $P4_12_12$	a=4.979Å Z=4 c=6.8903Å	Si(4a) O(8b)	$3\infty[Si^tO_2{}^{[2]*c*}]^c$ CRISTOBALITE	SR,52A,320;Wyckoff,1,316-317; LF,256.

Table 12

AB_2(cont.)

NAME	CHEMICAL FORMULA	STRUCTURAL FORMULA	SPACE GROUP	UNIT CELL DIMENSIONS	EQUIVALENT POSITIONS	STRUCTURE TYPE	REFERENCES
CUPRITE	Cu_2O	$3\infty[Cu_2^{[2]}O^{t}]$	Cub. $Pn\bar{3}m$	a=4.2696Å Z=2	Cu(4b) O(2a)	$3\infty[Cu_2^{[2]}O^{t}]$ CUPRITE	Zeit.Krist.,1988,182,158-160;LF, 272;Kostov,258;Wyckoff,1,331-332.
CUPROSTIBITE	$Cu_2(Sb,Tl)$	$Cu^{o}Cu^{t}[Sb,Tl]^{c}$	Tet. P4/nmm	a=3.99Å Z=2 c=6.09Å	Cu_{I}(2a)(Sb,Tl)(2c) Cu_{II}(2c)		Am.Min.,1970,55,1810-1811; Bokii,389.
DANBAITE	$CuZn_2$		Cub.	a=7.7615Å Z=11			Am.Min.,1984,69,566(Abs.); Hölzel,2.
DIGENITE	Cu_2S or Cu_9S_5	$Cu_2^{t/tr}[S]^{c}$ or $(Cu_9,\square)^{t/tr}[S_5]^{c}$	Trig. $R\bar{3}m$	a=3.92Å a_R=16.16Å c=48.0Å α=13°56' Z=3 Z_R=1(Cu_9S_5)		$Cu_2^{t/tr}[S]^{c}$ or $(Cu_9,\square)^{t/tr}[S_5]^{c}$ DIGENITE	Am.Min.,1963,48,110-123;Am.Min., 1958,43,228-242;LF,109;RRW,174; Hölzel;Pearson,1967,65,230.
DOWNEYITE	SeO_2		Tet. P4/mbc	a=8.36Å Z=8 c=5.07Å	Se(8h) O_{II}(8h) O_{I}(8g)	$1\infty[SeO_2]$ DOWNEYITE	Am.Min.,1977,62,316-320; Wyckoff,1,382-383;Pov.,299.
DRYSDALLITE	$Mo(Se,S)_2$	$Mo^{p}[(Se,S)_2]^{(2l8)h}$	Hex. $P6_3/mmc$	a=3.286Å Z=2 c=12.889Å		$Mo^{p}[S_2]^{(2l8)h}$ MOLYBDENITE -2H	Am.Min.,1974,59,1139(Abs.);Hölzel, 35;LF,154.
ERLICHMANITE	OsS_2	$Os^{o}\{g\}[S_2]^{c}$	Cub. $Pa\bar{3}$	a=5.6196Å Z=4		$Fe^{o}\{g\}[S_2]^{c}$ PYRITE	Am.Min.,1971,56,15011506;Pov., 736;RRW,195;Wyckoff,1,348.
FERROSELITE	$FeSe_2$	$Fe^{o}\{g\}[Se_2^{«h»}]^{c}$	Orth. Pnnm	a=4.8001Å Z=2 b=5.7706Å c=3.5850Å	Fe(2a) Se(4g)	$Fe^{o}\{g\}[S_2^{«h»}]^{c}$ MARCASITE	RRW,212;Pov.,243;Wyckoff,1,355-356;LF,169.
FLUORITE	CaF_2	$Ca^{cb}[F_2]^{t8}$	Cub. $Fm\bar{3}m$	a=5.46295Å (28°C) Z=4	Ca(4a) F(8c)	$Ca^{cb}[F_2]^{t8}$ FLUORITE	Wyckoff,1,239-244;LF,138.
FRANKDICKSONITE	BaF_2	$Ba^{cb}[F_2]^{t8}$	Cub. $Fm\bar{3}m$	a=6.1964Å Z=4	Ba(4a) F(8c)	$Ca^{cb}[F_2]^{t8}$ FLUORITE	Am.Min.,1974,59,885-888;Wyckoff, 1,240-241.
FROHBERGITE	$FeTe_2$	$Fe^{o}\{g\}[Te_2^{«h»}]^{c}$	Orth. Pnnm	a=5.268Å Z=2 b=6.273Å c=3.879Å	Fe(2a) Te(4g)	$Fe^{o}\{g\}[S_2^{«h»}]^{c}$ MARCASITE	SR,54A,71;Pov.,213;Am.Min.,1949, 34,342-382.
FROODITE	$PdBi_2$ α-form		Mon. C2/m	a=12.74Å β=102°27' b=4.29Å Z=4 c=5.71Å			Can.Min.,1958,6,200-209; SR,22,52-53.
GEVERSITE	$PtSb_2$	$Pt^{o}\{g\}[Sb_2]^{c}$	Cub. $Pa\bar{3}$	a=6.428Å Z=4		$Fe^{o}\{g\}[S_2]^{c}$ PYRITE	Min.Mag.,1961,32,833-847; Pov.,206;LF,171.
HASTITE	$CoSe_2$	$Co^{o}\{g\}[Se_2^{«h»}]^{c}$	Orth. Pnnm	a=3.60Å Z=2 b=4.84Å c=5.72Å		$Fe^{o}\{g\}[S_2^{«h»}]^{c}$ MARCASITE	Str. Tab.,135;Pov.,243;Am.Min., 1956,41,164-165;LF,169.
HAUERITE	MnS_2	$Mn^{o}\{g\}[S_2]^{c}$	Cub. $Pa\bar{3}$	a=6.092Å Z=4	Mn(4a) S(8c)	$Fe^{o}\{g\}[S_2]^{c}$ PYRITE	Min.Abs.,70-1595;Wyckoff,1, 346-348;Pov.,243;LF,171.
HEDLEYITE	$Bi_{2+x}Te_{1-x}$	$Bi^{o}[Bi_{1+x},Te_{1-x}]^{o/h}$	Trig. $R\bar{3}m$	a=4.47Å a_R=39.76Å c=119.04Å α=6°27' Z=18 Z_R=6			Pov.,214-215;Str.Tab.,131; Am.Min.,1949,34,342-382; Hölzel,30.
HESSITE	Ag_2Te	$Ag^{[5]}Ag^{t}[Te]^{c}$	Mon. $P2_1/c$	a=8.09Å β=123°20' b=4.48Å Z=4 c=8.96Å	Ag_{I-II}(4e) Te(4e)		Wyckoff,1,340-341;Ross,763;Pov., 211;LF,262.
ICE	H_2O	$3\infty[O^{t}H_2^{[2]«ch»}]^{h}$	Hex. $P6_3/mmc$	a=4.5227Å Z=4 (0°C) c=7.367Å	O(4f)	$3\infty[O^{t}H_2^{[2]«ch»}]^{h}$ ICE	Wyckoff,1,322-324;Hölzel,58, LF,262.

Table 13

AB_2(cont.)

NAME	CHEMICAL FORMULA	STRUCTURAL FORMULA	SPACE GROUP	UNIT CELL DIMENSIONS	EQUIVALENT POSITIONS	STRUCTURE TYPE	REFERENCES
INSIZWAITE	$Pt(Bi,Sb)_2$	$Pt^{o}\{g\}[(Bi,Sb)_2]^{c}$	Cub. $Pa\bar{3}$	a=6.625Å Z=4		$Fe^{o}\{g\}[S_2]^{c}$ PYRITE	Min.Mag.,1972,38,794-800; Wyckoff,1,348.
IRIDARSENITE	$(Ir,Ru)As_2$	$(Ir,Ru)^{o}\{g\}[As_2^{«n»}]^{c}$	Mon.	a=6.05Å β=113°17' b=6.06Å Z=4 c=6.18Å			Can.Min.,1974,12,280-284; K/S,160;Pearson,1972,284.
IXIOLITE	$(Ta,Mn,Nb)O_2$	$(Ta,Mn,Nb)^{o}[O_2]^{n}$	Orth. Pbcn	a=4.785Å Z=4 b=5.758Å c=5.160Å	(Ta,Mn,Nb)(4c) O(8d)	$Pb^{o}[O_2]^{n}$ SCRUTINYITE	Can.Min.,1976,14,540-549; Pov.,302-303;LF,92.
KHATYRKITE	$(Cu,Zn)Al_2$	$(Cu,Zn)^{acb}[Al_2]^{N2 1 1}$	Tet. I 4/mcm	a=6.066Å Z=4 c=4.874Å	Cu(4a) Al(8h)	$Au^{acb}[Pb_2]^{N2 1 1}$ ANYUIITE	ZVMO,1985,114,90;Pearson, 1967,111,608;Bokii,330-331.
KRENNERITE	$AuTe_2$	$Au^{o}[Te_2]^{n}$	Orth. Pma2	a=16.58Å Z=8 b=8.849Å c=4.464Å	Au_{I}(2a) Te_{I-II}(2c) Au_{II}(2c) Te_{III-V}(4d) Au_{III}(4d)	Dist.deriv. $Ni^{o}[Te_2]^{n}$ MELONITE	Min.Abs.,85M/2407;K/S,22-23;Bokii,404;Pov.,212.
KRUTAITE	$CuSe_2$	$Cu^{o}\{g\}[Se_2]^{c}$	Cub. $Pa\bar{3}$	a=6.056Å Z=4		$Fe^{o}\{g\}[S_2]^{c}$ PYRITE	Am.Min.,1974,59,210(Abs.). Hölzel,32.
KRUTOVITE	$NiAs_2$	$Ni^{o}\{g\}[As_2]^{c}$	Cub. $P2_13$	a=5.794Å Z=4		Dist.deriv. $Fe^{o}\{g\}[S_2]^{c}$ PYRITE	Am.Min.,1977,62,173-174 (Abs.);K/S,32,164;Hölzel,33.
KULLERUDITE	$NiSe_2$	$Ni^{o}\{g\}[Se_2^{«n»}]^{c}$	Orth. Pnnm	a=4.89Å Z=2 b=5.96Å c=3.67Å		$Fe^{o}\{g\}[S_2^{«n»}]^{c}$ MARCASITE	Am.Min.,1965,50,519-520; Pov.,243;Str.Tab.,135.
LAURELITE	$Pb(F,Cl,OH)_2$		Hex. P6...	a=10.252Å Z=6 c=3.973Å			Am.Min.,1989,74,927-933; Hölzel,suppl..
LAURITE	RuS_2	$Ru\{g\}[S_2]^{c}$	Cub. $Pa\bar{3}$	a=5.6089Å Z=4	Ru(4a) S(8c)	$Fe^{o}\{g\}[S_2]^{c}$ PYRITE	Min.Mag.,1983,47,465-471; Wyckoff,1,348;Str.Tab.,132.
LAWRENCITE	$(Fe,Ni)Cl_2$	$(Fe,Ni)^{o}[Cl_2]^{c}$	Trig. $R\bar{3}m$	a=3.5775Å a_R=6.19Å c=17.52Å α=33°33' Z=3 Z_R=1		$Mg^{o}[Cl_2]^{c}$ CHLOROMA-GNESITE	RRW,347;Wyckoff,1,270-272;Pov.644;LF,97.
LECHATELIERITE	SiO_2	---	Amorf	---	---	---	RRW,350;Str.Tab.,195.
LÖLLINGITE	$FeAs_2$	$Fe^{o}\{g\}[As_2^{«n»}]^{c}$	Orth. Pnnm	a=5.3001Å Z=2 b=5.9838Å c=2.8821Å	Fe(2a) As(4g)	$Fe^{o}\{g\}[S_2^{«n»}]^{c}$ MARCASITE	SR,54A,43;Hölzel,34;Am. Min.,1968,53,1856-1881; Wyckoff,1,356.
MALDONITE	Au_2Bi	$[[Au_2][Bi]]^{(-TK)c}$	Cub. $Fd\bar{3}m$	a=7.98Å Z=8	Bi(8a) Au(16d)	$[[Au_2][Bi]]^{(-TK)c}$ MALDONITE	Pov.,204;Bokii,386,316-317; Wyckoff,1,365-367.
MARCASITE	FeS_2	$Fe\{g\}[S_2^{«n»}]^{c}$	Orth. Pnnm	a=4.436Å Z=2 b=5.414Å c=3.381Å	Fe(2a) S(4g)	$Fe\{g\}[S_2^{«n»}]^{c}$ MARCASITE	Wyckoff,1,355;Str.Tab.,135; SR,35A,75-76;LF,169.
MATTAGAMITE	$(Co,Fe)Te_2$	$(Co,Fe)^{o}\{g\}[Te_2^{«n»}]^{c}$	Orth. Pmnn	a=5.305Å Z=2 b=6.289Å c=3.866Å		$Fe^{o}\{g\}[S_2^{«n»}]^{c}$ MARCASITE	Am.Min.,1974,59,382-383 (Abs.);Hölzel,33;SR,35A, 75;Wyckoff,1,356.
MCKINSTRYITE	$(Ag,Cu)_2S$		Orth. Pnam...	a=14.043Å Z=32 b=15.677Å c=7.803Å			Econ.Geol.,1966,61,1383-1389;Hölzel,13.

Table 14

AB_2(cont.)

NAME	CHEMICAL FORMULA	STRUCTURAL FORMULA	SPACE GROUP	UNIT CELL DIMENSIONS	EQUIVALENT POSITIONS	STRUCTURE TYPE	REFERENCES
MELONITE	$NiTe_2$	$Ni^{o}[Te_2]^{n}$	Trig. P$\bar{3}$m1	a=3.861Å c=5.297Å Z=1	Ni(1a) Te(2d)	$Ni^{o}[Te_2]^{n}$ MELONITE	Wyckoff,1,267-269;Pov.,214; LF,98.
MERENSKYITE	$(Pd,Pt)(Te,Bi)_2$	$(Pd,Pt)^{o}[(Te,Bi)_2]^{n}$	Trig. P$\bar{3}$m1	a=3.978Å c=5.125Å Z=1		$Ni^{o}[Te_2]^{n}$ MELONITE	Min.Mag.,1966,35,815-834; Pov.,214;Str.Tab.,138;LF,98.
MOLYBDENITE-2H	MoS_2	$Mo^{p}[S_2]^{(2Ts)n}$	Hex. $P6_3/mmc$	a=3.1604Å (26°C) c=12.295Å Z=2	Mo(2c) S(4f)	$Mo^{p}[S_2]^{(2Ts)n}$ MOLYBDENITE-2H	Wyckoff,1,280-282;Pov.,259; LF,154.
MOLYBDENITE-3R	MoS_2	$Mo^{p}[S_2]^{(2Ts)c}$	Trig. R$\bar{3}$m	a=3.16Å c=18.45Å Z=3	Mo(3a) S_{I-II}(3a)	$Mo^{p}[S_2]^{(2Ts)c}$ MOLYBDENITE-3R	Wyckoff,1,280-282;RRW,412.
MONCHEITE	$(Pt,Pd)(Te,Bi)_2$	$(Pt,Pd)^{o}[(Te,Bi)_2]^{n}$	Trig. P$\bar{3}$m1	a=4.049Å c=5.288Å Z=1		$Ni^{o}[Te_2]^{n}$ MELONITE	Am.Min.,1963,48,1181(Abs.); Pov.,214.
NAUMANNITE	Ag_2Se	$3\infty[Ag_2{}^{[2]}Se^{t}]$	Orth. $P2_12_12_1$	a=4.333Å Z=4 b=7.062Å c=7.764Å	Ag_{I-II}(4a) Se(4a)	Dist.deriv. $3\infty[Cu_2{}^{[2]}O^{t}]$ CUPRITE	Am.Min.,1971,56,1882-1888; Pov.,241;Sov.Phys.Cryst.,1965, 10,225-231.
NISBITE	$NiSb_2$	$Ni^{o}\{g\}[Sb_2{}^{«h»}]^{c}$	Orth. Pnnm	a=5.162Å Z=2 b=6.303Å c=3.839Å		$Fe^{o}\{g\}[S_2{}^{«h»}]^{c}$ MARCASITE	Am.Min.,1971,56,631-632(Abs.); Pov.,206;LF,169.
NSUTITE	$Mn(O,OH)_2$	$Mn^{o}[(O,OH)_2]^{n}$	Hex.	a=9.65Å c=4.43Å Z=12			Am.Min.,1965,50,170-179;Am. Min.,1979,64,1199-1218;RRW, 444;Acta Cryst.,1959,12,341-345.
PALLADOARSENIDE	Pd_2As		Mon. P2/m	a=9.25Å β=94° b=8.47Å Z=18 c=10.44Å			Am.Min.,1975,60,162(Abs.);K/S, 158;Hölzel,11.
PALLADOBISMUTH-ARSENIDE	$Pd_2(As,Bi)$		Orth. Pmcn...	a=7.504Å Z=20 b=18.884Å c=6.841Å			Can.Min.,1976,14,410-411;K/S, 158.
PAOLOVITE	Pd_2Sn		Orth. Pbnm	a=8.11Å Z=4 b=5.662Å c=4.324Å			Am.Min.,1974,59,1331(Abs.); Hölzel,6.
PARAMONTROSEI-TE	VO_2	$V^{o}[O_2]^{n}$	Orth. Pbnm	a=4.89Å Z=4 b=9.39Å c=2.93Å	V(4c) O_{I-II}(4c)	$Mn^{o}[O_2]^{n}$ RAMSDELLITE	Wyckoff,1,290-293;Pov.,298; Hölzel,83.
PARARAMMELS-BERGITE	$NiAs_2$	$Ni^{o}\{g\}[As_2{}^{«h»}]^{c}$	Orth. Pbca	a=5.753Å Z=8 b=5.799Å c=11.407Å	Ni(8c) As_{I-II}(8c)	Dist.deriv. $Fe^{o}\{g\}[S_2{}^{«h»}]^{c}$ MARCASITE	Am.Min.,1972,57,1-9;K/S,34-35; SR,38A,27-28.
PARATELLURITE	TeO_2		Tet. $P4_32_12$	a=4.810Å c=7.613Å Z=4			Sov.Phy.Cryst.,1987,32,354-359;Am.Min.,1960,45,1272-1274;Pov.,298-299.

Table 15

AB_2(cont.)

NAME	CHEMICAL FORMULA	STRUCTURAL FORMULA	SPACE GROUP	UNIT CELL DIMENSIONS	EQUIVALENT POSITIONS	STRUCTURE TYPE	REFERENCES
PAXITE	$CuAs_2$		Mon. $P2_1/c$	a=5.83Å β=99°.7 b=5.11Å Z=? c=8.08Å			TMPM,1985,34,167-?;Pov.,204; Str.Tab.,129.
PENROSEITE	$NiSe_2$	$Ni^o\{g\}[Se_2]^c$	Cub. $Pa\bar{3}$	a=5.9629Å Z=4		$Fe^o\{g\}[S_2]^c$ PYRITE	Am.Min.,1989,74,1168-1176;Pov., 243;LF,171.
PLATTNERITE	PbO_2	$Pb^o[O_2]^n$	Tet. $P4_2/mnm$	a=4.946Å Z=2 c=3.379Å		$Ti^o[O_2]^n$ RUTILE	Wyckoff,1,250-251;Pov.,297;LF,88.
PORTLANDITE	$Ca(OH)_2$	$Ca^o[(OH)_2]^n$	Trig. $P\bar{3}m1$	a=3.5844Å Z=1 c=4.8962Å	Ca(1a) (OH)(2d)	$Ni^o[Te_2]^n$ MELONITE	Wyckoff,1,266-269;Pov.,324;LF,98.
PYRITE	FeS_2	$Fe^o\{g\}[S_2]^c$	Cub. $Pa\bar{3}$	a=5.40667Å Z=4	Fe(4a) S(8c)	$Fe^o\{g\}[S_2]^c$ PYRITE	Wyckoff,1,346-348;Pov.,243; Am.Min.,1989,74,1168-1176;LF,171.
PYROCHROITE	$Mn(OH)_2$	$Mn^o[(OH)_2]^n$	Trig. $P\bar{3}m1$	a=3.323Å Z=1 c=4.738Å		$Ni^o[Te_2]^n$ MELONITE	Am.Min.,1965,50,1296-1354;Pov.,324; Wyckoff,1,267-268;Str.Tab.,214-215..
PYROLUSITE	MnO_2 β-form	$Mn^o[O_2]^n$	Tet. $P4_2/mnm$	a=4.3983Å Z=2 c=2.8730Å	Mn(2a) O(4f)	$Ti^o[O_2]^n$ RUTILE	Acta Cryst.,1976,B32,2200-2004; Pov.,297;LF,88.
QUARTZ	SiO_2 α-form	$3\infty[Si^tO_2{}^{eIVs}]$	Trig. $P3_121$	a=4.9134Å Z=3 c=5.4052Å	Si(3a) O(6c)	$3\infty[Si^tO_2{}^{eIVs}]$ QUARTZ	Acta Cryst.,1976,B32,2456-2459; Pov.,291;LF,264.
RAMMELSBER-GITE	$NiAs_2$	$Ni^o\{g\}[As_2{}^{ehs}]^c$	Orth. Pnnm	a=4.78Å b=5.78Å c=3.53Å	Ni(2a) As(4g)	$Fe^o\{g\}[S_2{}^{ehs}]^c$ MARCASITE	Wyckoff,1,355-356;Min.Abs.,771487; Pov.,206;LF,169.
RAMSDELLITE	MnO_2 γ-form	$Mn^o[O_2]^n$	Orth. Pbnm	a=4.533Å Z=4 b=9.27Å c=2.866Å	Mn(4c) $O_{I\text{-}II}$(4c)	$Mn^o[O_2]^n$ RAMSDELLITE	Wyckoff,1,290-293;Am.Min.,1979,64, 1199-1218:LF,94.
ROXBYITE	$Cu_{1.78}S$	$(Cu_{1.78},\square_{0.22})^{trt}[S]^h$	Mon. C2/m...	a=53.79Å β=90°.0 b=30.90Å Z=64 c=13.36Å			Min.Mag.,1988,52,323-330.
RUTILE	TiO_2	$Ti^o[O_2]^n$	Tet. $P4_2/mnm$	a=4.59373Å (25°C) c=2.95812Å Z=2	Ti(2a) O(4f)	$Ti^o[O_2]^n$ RUTILE	Wyckoff,1,250-251;Pov.,297;LF,88.
SAFFLORITE	$CoAs_2$	$Co^o\{g\}[As_2{}^{ehs}]^c$	Orth. Pnnm	a=5.81Å Z=2 b=6.36Å c=4.87Å		$Fe^o\{g\}[S_2{}^{ehs}]^c$ MARCASITE	Am.Min.,1968,53,1856-1881;RRW, 531;Pov.,206;Hölzel,34;LF,169.
SCACCHITE	$MnCl_2$	$Mn^o[Cl_2]^c$	Trig. $R\bar{3}m$	a=3.686Å a_R=6.20Å c=17.470Å α=34°35' Z=3 Z_R=1		$Mg^o[Cl_2]^c$ CHLOROMAGNESITE	Wyckoff,1,270-272;Pov.,644;LF,97.
SCRUTINYITE	PbO_2 α-form	$Pb^o[O_2]^n$	Orth. Pbcn	a=4.937Å Z=4 b=5.939Å c=5.486Å	Pb(4c) O(8d)	$Pb^o[O_2]^n$ SCRUTINYITE	SR,16,224-225;Can.Min.,1988, 26,905-910;Hölzel,suppl..
SEINÄJOKITE	(Fe,Ni) $(Sb,As)_2$	$(Fe,Ni)^o$ $\{g\}_c[(Sb,As)_2{}^{ehs}]$	Orth. Pnnm	a=3.19Å Z=2 b=5.81Å c=6.49Å	(Fe,Ni)(2a) (Sb,As)(4g)	$Fe^o\{g\}[S_2{}^{ehs}]^c$ MARCASITE	Am.Min.,1977,62,1059-1060(Abs.); Garcia Orta,Sér.Geol.,1986,9,31-40.
SELLAITE	MgF_2	$Mg^o[F_2]^n$	Tet. $P4_2/mnm$	a=4.623Å Z=2 c=3.052Å	Mg(2a) F(4f)	$Ti^o[O_2]^n$ RUTILE	Wyckoff,1,250-258;Pov.,666;LF,88.

Table 16

AB_2(cont.)

NAME	CHEMICAL FORMULA	STRUCTURAL FORMULA	SPACE GROUP	UNIT CELL DIMENSIONS	EQUIVALENT POSITIONS	STRUCTURE TYPE	REFERENCES
SPERRYLITE	$PtAs_2$	$Pt^{o}\{g\}[As_2]^{c}$	Cub. $Pa\bar{3}$	a=5.9861Å Z=4	Pt(4a) As(8c)	$Fe^{o}\{g\}[S_2]^{c}$ PYRITE	SR,45A,27;Pov.,206;Can. Min.,1979,17,117-123;LF,171.
SPERTINIITE	$Cu(OH)_2$	$Cu^{o}[(OH)_2]^{R31\ my}$	Orth. Bbmm	a=10.59Å Z=4 b=5.256Å c=2.949Å	Cu(4c) $(OH)_{I-II}$(4c)	$Fe^{o}[O(OH)]^{R31\ my}$ LEPIDOCROCITE	Wyckoff,1,293-295;Hölzel,82.
SRILANKITE	$(Ti,Zr)O_2$	$(Ti,Zr)^{o}[O_2]^{n}$	Orth. Pbcn	a=4.706Å Z=4 b=5.553Å c=5.024Å	(Ti,Zr)(4c) O(8d)	$Pb^{o}[O_2]^{n}$ SCRUTINYITE	Zeit.Krist.,1983,164,59-66; Am.Min.,1984,69,212(Abs.); Hölzel,75.
STISHOVITE	SiO_2	$Si^{o}[O_2]^{n}$	Tet. $P4_2/mnm$	a=4.1790Å Z=2 c=2.6649Å		$Ti^{o}[O_2]^{n}$ RUTILE	Am.Min.,1962,47,807(Abs.); Hölzel,71;LF,88.
SWEETITE	$Zn(OH)_2$		Tet. $P4_12_12$...	a=8.222Å c=14.34Å Z=20			Min.Mag.,1984,48,267-269; Am.Min.,1985,70,438(Abs.); Hölzel,82.
TELLURITE	TeO_2	$2\infty[Te^{[4]}O_2]$	Orth. Pbcc	a=12.035Å Z=8 b=5.464Å c=5.607Å	Te(8c) O_{I-II}(8c)	$2\infty[Te^{[4]}O_2]$ TELLURITE	Zeit.Krist.,1967,124,228-237; Pov.,309;Hölzel,75.
THEOPHRASTITE	$Ni(OH)_2$	$Ni^{o}[(OH)_2]^{n}$	Trig. $P\bar{3}m1$	a=3.131Å Z=1 c=4.608Å		$Ni^{o}[Te_2]^{n}$ MELONITE	Am.Min.,1981,66,1020-1021; Wyckoff,1,268;Hölzel,52.
THORIANITE	ThO_2	$Th^{cb}[O_2]^{Q8}$	Cub. $Fm\bar{3}m$	a=5.5997Å Z=4	Th(4a) O(8c)	$Ca^{cb}[F_2]^{Q8}$ FLUORITE	Wyckoff,1,239-243;Pov.,269; Str.Tab.,211;LF,138.
TOLBACHITE	$CuCl_2$	$Cu^{o}[Cl_2]^{c}$	Mon. C2/m	a=6.9038Å β=121.197° b=3.2995Å Z=2 c=6.824Å	Cu(2b) Cl(4i)	Dist.der. $Mg^{o}[Cl_2]^{c}$ CHLOROMAGNE-SITE	Am.Min.,1993,78,187-189; Wyckoff,1,345-346;Am.Min., 1984,69,408(Abs.).
TRIDYMITE	SiO_2 α-form	$3\infty[Si^{t}O_2^{[2]*ch*}]^{h}$	Mon. Cc	a=18.494Å β=117.75° b=4.991Å Z=48 c=25.832Å	Si_{I-XII}(4a) O_{I-XXIV}(4a)	$3\infty[Si^{t}O_2^{[2]*ch*}]^{h}$ TRIDYMITE	Acta Cryst.,1976,B32,2486-2491;Pov.,291-292;LF,259.
TROGTALITE	$CoSe_2$	$Co^{o}\{g\}[Se_2]^{c}$	Cub. $Pa\bar{3}$	a=5.8588Å Z=4	Co(4a) Se(8c)	$Fe^{o}\{g\}[S_2]^{c}$ PYRITE	SR,19,123;Pov.,243;LF,171.
TUGARINOVITE	MoO_2	$Mo^{o}[O_2]^{n}$	Mon. $P2_1/c$	a=5.59Å β=119° 32' b=4.82Å Z=4 c=5.51Å	Mo(4e) O_{I-II}(4e)	Dist.deriv. $Ti^{o}[O_2]^{n}$ RUTILE	Am.Min.,1981,66,438-439 (Abs.);Wyckoff,1,257-259; Hölzel,80.
TUNGSTENITE-2H	WS_2	$W^{p}[S_2]^{(2T s)h}$	Hex. $P6_3/mmc$	a=3.154Å Z=2 c=12.362Å	W(2c) S(4f)	$Mo^{p}[S_2]^{(2T s)h}$ MOLYBDENITE-2H	RRW,633;Wyckoff,1,280-281;Pov.,259;LF,154.
TUNGSTENITE-3R	WS_2	$W^{p}[S_2]^{(2T s)c}$	Trig. $R\bar{3}m$	a=3.16Å Z=3 c=18.45Å	W(3a) S_{I-II}(3a)	$Mo^{p}[S_2]^{(2T s)c}$ MOLYBDENITE-3R	Wyckoff,1,280-282;Hölzel,35; JCPDS,35-651.
URANINITE	UO_2	$U^{cb}[O_2]^{Q8}$	Cub. $Fm\bar{3}m$	a=5.4682Å (26°C) Z=4	U(4a) O(8c)	$Ca^{cb}[F_2]^{Q8}$ FLUORITE	Wyckoff,1,239-243;Pov.,273; RRW,640;LF,138.
URVANTSEVITE	$Pd(Bi,Pb)_2$		Hex.	a=13.82Å Z=12 ? c=6.53Å			Am.Min.,1977,62,1260-1261 (Abs.);Hölzel,33.
VAESITE	NiS_2	$Ni^{o}\{g\}[S_2]^{c}$	Cub. $Pa\bar{3}$	a=5.670Å Z=4	Ni(4a) S(8c)	$Fe^{o}\{g\}[S_2]^{c}$ PYRITE	Am.Min.,1989,74,1168-1176: SR,24,231-232;Pov.,243; Str.Tab.,132.

Table 17

AB_2(cont.)

NAME	CHEMICAL FORMULA	STRUCTURAL FORMULA	SPACE GROUP	UNIT CELL DIMENSIONS	EQUIVALENT POSITIONS	STRUCTURE TYPE	REFERENCES
VILLAMANINITE	$(Cu,Ni,Co,Fe)S_2$	$(Cu,Ni,Co,Fe)^{o}$ $\{g\}[S_2]^{c}$	Cub. $Pa\bar{3}$	a=5.7900Å Z=4		$Fe^{o}\{g\}[S_2]^{c}$ PYRITE	RRW,652;Am.Min.,1989,74,1168-1176;Pov.,243;LF,171.
WEISSITE	$Cu_{2-x}Te$		Hex. P6/mmm	a=12.45Å Z=54 c=21.56Å			RRW,666;Hölzel,13;Str.Tab.,108;Pov.,211-212.
WÜLFINGITE	$Zn(OH)_2$	$_{3\infty}[Zn^{t}(OH)_2]$	Orth. $P2_12_12_1$	a=8.490Å Z=4 b=5.162Å c=4.917Å	Zn(4a) O_{I-II}(4a)		Am.Min.,1988,73,196-197 (Abs.);Wyckoff,1,325-326.
ZIRKELITE	$(Ti,Ca,Zr)O_{2-x}$	$(Ti,Ca,Zr)^{cb}$ $[O_{2-x}]^{Qs}$	Cub.	a=5.02Å Z=4 ?			Min.Mag.,1989,53,365-369;Am. Min.,1983,68,262-276;Am.Min.,1982,67,615-620;Hölzel,68.

Table 18

AB_3

NAME	CHEMICAL FORMULA	STRUCTURAL FORMULA	SPACE GROUP	UNIT CELL DIMENSIONS	EQUIVALENT POSITIONS	STRUCTURE TYPE	REFERENCES
ATHENEITE	$(Pd,Hg)_3As$		Hex. P6/mmm	a=6.798Å Z=2 c=3.483Å			Min.Mag.,1974,39,528-543; Hölzel,10.
ATOKITE	$(Pd,Pt)_3Sn$	$[(Pd,Pt)_3Sn]^c$	Cub. $Fm\bar{3}m$	a=3.991Å Z=4			Can.Min.,1975,13,146-150; Bokii,294-295.
AURICUPRIDE	Cu_3Au	$[Cu_3Au]^c$	Cub. $Pm\bar{3}m$	a=3.75Å Z=1	Au(1a) Cu(3c)	$[Cu_3Au]^c$ AURICUPRIDE	Pov.,194;Megaw,90;ZVMO. 1977,106,540.
AWARUITE	Ni_3Fe	$[Ni_3Fe]^c$	Cub. $Pm\bar{3}m$	a=3.59Å Z=1		$[Cu_3Au]^c$ AURICUPRIDE	Str.Tab.,96;Can.Min.,1959,6, 307-319;Hölzel,3.
BAYERITE	$Al(OH)_3$ α-form	$Al^o[(OH)_3]^{(2Ts)h}$	Mon. $P2_1/n$	a=5.0626Å β=90.26° b=8.6719Å Z=8 c=9.4254Å	$Al_{I-II}(4e)$ $O_{I-VI}(4e)$ $H_{I-VI}(4e)$	Dist.deriv. $Al^o[(OH)_3]^{(2Ts)h}$ GIBBSITE	Zeit.Krist.,1978.148,255-273; SR,45A,271;Hölzel,82;Pov., 323;Str.Tab.,214;LF,155.
COHENITE	Fe_3C	$C^p[Fe_3]^{1s}$	Orth. Pnma	a=5.09Å Z=4 b=6.748Å c=4.523Å	C(4c) $Fe_I(4c)$ $Fe_{II}(8d)$		Acta Cryst.,1964,17,1331-1332; Pov,201;SR,29,34-35;Wyckoff, 2,112-115.
DOMEYKITE	Cu_3As α-form	$Cu_3^t[As]^b$	Cub. $I\bar{4}3d$	a=9.619Å Z=16	Cu(48e) As(16c)		Zeit.Krist.,1977,145,334-345; SR,44A,22;Pov.,205.
DOYLEITE	$Al(OH)_3$	$Al^o[(OH)_3]^{c/h}$	Tric. P1	a=5.002Å α=97.50° b=5.175Å β=118.60° c=4.980Å γ=104.74° Z=2			Can.Min.,1985,23,21-28; Hölzel.82;Am.Min.,1986,71,845 (Abs.).
DYSCRASITE	Ag_3Sb	$[Ag_3Sb]^h$	Orth. Pmm2	a=3.008Å Z=1 b=4.828Å c=5.214Å	Sb(1a) $Ag_{II}(1b)$ $Ag_I(1c)$ $Ag_{III}(1d)$	$[Ag_3Sb]^h$ DYSCRASITE	Can.Min.,1976,14,139-142; Pov.,205;SR,43A,17.
DZHALINDITE	$In(OH)_3$	$In^o[\square(OH)_3]^c$	Cub. $Im\bar{3}$	a=7.95Å Z=8	In(8c) (OH)(24g)	Defect deriv. $Ti^o[CaO_3]^c$ PEROVSKITE	Am.Min.,1964,49,439-440 (Abs.);Wyckoff,2,76;Pov.,315.
FLUOCERITE-(Ce)	$(Ce,La)F_3$	$(Ce,La)^{tr}[F_3]^{cch}$	Hex. $P6_3/mcm$	a=7.130Å Z=6 c=7.298Å			Min.Mag.,1983,47,41-46; Pov.,667.
FLUOCERITE-(La)	$(La,Ce)F_3$	$(La,Ce)^{tr}[F_3]^{cch}$	Trig. $P\bar{3}c1$	a=4.135Å Z=6 c=7.295Å	La(6g) $F_{II}(4c)$ $F_I(2a)$ $F_{III}(12k)$		Am.Min.,1984,69,566(Abs.); Hölzel,49;Bokii,407-408;Acta Cryst.,1985,B41,81-93.
GANANITE	BiF_3 α-form	$Bi^o[F_3]^c$	Cub. $P\bar{4}3m$	a=5.825Å Z=4			Am.Min.,1988,73,1494(Abs.); Bokii,456;Hölzel,49;Wells,1962, 342-343.
GIBBSITE	$Al(OH)_3$ γ-form	$Al^o[(OH)_3]^{(2Ts)h}$	Mon. $P2_1/n$	a=8.684Å β=94.54° b=5.078Å Z=8 c=9.736Å	$Al_{I-II}(4e)$ $H_{I-VI}(4e)$ $O_{I-VI}(4e)$	$Al^o[(OH)_3]^{(2Ts)h}$ GIBBSITE	Zeit.Krist.,1974,139,129-135; Wyckoff,2,78-80;Pov.,323-324; Wells,1962,552-553;LF,155.
GUPEIITE	Fe_3Si	$[Fe_3Si]^b$	Cub. $Fm\bar{3}m$	a=5.670Å Z=4	Si(4a) $Fe_{II}(8c)$ $Fe_I(4b)$		Am.Min.,1986,71,228(Abs.);Bokii, 308-309,456;Wells,1962,997.
ISOFERROPLATINUM	$(Pt,Pd)_3$ (Fe,Cu)		Cub. $Pm\bar{3}m$...	a=3.86Å Z=1			Hölzel,6;Can.Min.,1975,13,117- 126.
LUANHEITE	Ag_3Hg		Hex.	a=6.61Å Z=6 c=10.98Å			Am.Min.,1988,73,192-193 (Abs.);Hölzel,2.

Table 19

AB_3(cont.)

NAME	CHEMICAL FORMULA	STRUCTURAL FORMULA	SPACE GROUP	UNIT CELL DIMENSIONS	EQUIVALENT POSITIONS	STRUCTURE TYPE	REFERENCES
MOLYBDITE	MoO_3	$Mo^o[O_3]^c$	Orth. Pbnm	a=3.954Å Z=4 b=13.808Å c=3.690Å	Mo(4c) O_{I-III}(4c)	$Mo^o[O_3]^c$ MOLYBDITE	Am.Min.,1988,<u>49</u>,1497-1498 (Abs.);Wyckoff,<u>2</u>,81;Pov.,307-308.
MOLYSITE	$FeCl_3$	$Fe^o[Cl_3]^h$	Trig. $R\bar{3}$	a=6.0650Å a_R=6.70Å c=17.420Å α=52°30′ Z=6 Z_R=2	Fe(6c) Cl(18f)	$Fe^o[Cl_3]^h$ MOLYSITE	J.Appl.Cryst.,1989,<u>22</u>,173-180;Wyckoff,<u>2</u>,45-46;Pov.,643-644.
NICKEL-SKUTTE-RUDITE	$NiAs_{2-3}$	$Ni_4{}^o\{g\}[As_4]_{2-3}$	Cub. $Im\bar{3}$	a=8.28Å Z=8			Str.Tab.,138;Hölzel,36;RRW, 435-436.
NORDSTRANDITE	$Al(OH)_3$	$Al^o[(OH_3]^{(21s)c}$	Tric. P1...	a=6.148Å α=95.76° b=6.936Å β=99.06° c=5.074Å γ=8330° Z=4			Can.Min.,1982,<u>20</u>,77-85; Can.Min.,1985,<u>23</u>,21-28; Pov.,223.
PARADOCRASITE	Sb_3As		Mon. C2	a=7.252Å β=123°0.8′ b=4.172Å Z=1 c=4.431Å			Am.Min.,1971,<u>56</u>,1127-1146; Hölzel,8.
RUSTENBURGITE	Pt_3Sn	$[Pt_3Sn]^c$	Cub. $Fm\bar{3}m$	a=3.991Å Z=4			Can.Min.,1975,<u>13</u>,146-150; Bokii,294-295.
SASSOLITE	$B(OH)_3$		Tric. P1	a=7.04Å α=92°35′ b=7.05Å β=101°10′ c=6.58Å γ=119°50′ Z=4			Pov.,486-487;Hölzel,82;Acta Cryst.,1954,<u>7</u>,305- .
SCHREIBERSITE	$(Fe,Ni,Cr)_3P$		Tet. $I\bar{4}$	a=9.040Å Z=8 c=4.462Å	P(8g) $(Fe,Ni,Cr)_{I-III}$(8g)		SR,<u>35A</u>,71-72;Wyckoff,<u>2</u>,103-105.
SIDERAZOT	$Fe_{2.5}N$	$N^o[Fe_{2.5}]^h$	Trig. $P\bar{3}1m$	a=2.74Å Z= ? c=4.39Å			Pov.,200-201;Zeit.Krist.,1930, <u>74</u>,511-533;Str.Tab.,97;Hölzel,5.
SKUTTERUDITE	$CoAs_3$	$Co_4{}^o\{g\}[As_4]_3$	Cub. $Im\bar{3}$	a=8.2055Å Z=8	Co(8c) As(24g)	$Co_4{}^o\{g\}[As_4]_3$ SKUTTERUDITE	SR,<u>40A</u>,6;Pov.,207;K/S,35;LF, 176.
SÖHNGEITE	$Ga(OH)_3$	$Ga^o[\square(OH)_3]^c$	Cub. $Im\bar{3}$	a=7.47Å Z=8		$In^o[\square(OH)_3]^c$ DZHALINDITE	Am.Min.,1966,<u>51</u>,1815(Abs.); Wells,1962,551;Pov.,315;Str. Tab.,219.
SUESSITE	Fe_3Si	$[Fe_3Si]^b$	Cub. $Fm\bar{3}m$	a=2.841Å Z=?			Am.Min.,1982,<u>67</u>,126-131;Wells, 1962,997;Hölzel,4.
TELARGPALITE	$(Pd,Ag)_3Te$		Cub.	a=12.60Å Z=4			K/S,158-159;Am.Min.,1981,<u>66</u>, 1103(Abs.);Hölzel,11.
VINCENTITE	$(Pd,Pt)_3$ (As,Sb,Te)		Tet.				Min.Mag.,1974,<u>39</u>,525-527; Hölzel,10;Encyc.Miner.Nam.321.
ZVYAGINTSEVITE	$Pd_3(Pb,Sn)$	$[Pd_3(Pb,Sn)]^c$	Cub. $Pm\bar{3}m$	a=4.021Å Z=1	(Pb,Sn)(1a) Pd(3c)	$[Cu_3Au]^c$ AURICUPRIDE	Can.Min.,1966,<u>8</u>,541-550;Pov., 194-195;Str.Tab.,99;Megaw,90.

Table 20

A_2B_3

NAME	CHEMICAL FORMULA	STRUCTURAL FORMULA	SPACE GROUP	UNIT CELL DIMENSIONS	EQUIVALENT POSITIONS	STRUCTURE TYPE	REFERENCES
ARSENOLITE	As_2O_3		Cub. $Fd\bar{3}m$	a=11.07Å Z=16	As(32e) O(48f)		SR,9,164-165;Pov.,297;Str.Tab., 182.
AVICENNITE	Tl_2O_3	$Tl_2^{[6]}[O_3\square]^{Cs}$	Cub. $Ia\bar{3}$	a=10.52Å Z=16		$Mn_2^{[6]}[O_3\square]^{Cs}$ BIXBYITE	Pov.,271;Str.Tab.,184;Hölzel, 63;LF,139.
BISMITE	Bi_2O_3 α-form		Mon. $P2_1/c$	a=5.83Å β=67°04' b=8.14Å Z=4 c=7.48Å			RRW,73;Pov.,271;Hölzel,63.
BISMUTHINITE	Bi_2S_3	$Bi_2^{[3+4]}[S_3]'$	Orth. Pbnm	a=11.115 Z=4 b=11.25Å c=3.97Å	$Bi_{I-II}(4c)$ $S_{I-III}(4c)$	$Sb_2^{[3+4]}[S_3]'$ STIBNITE	SR,43A,33-34;Pov.,246,128; LF,160.
BIXBYITE	Mn_2O_3	$Mn_2^{[6]}[O_3\square]^{Cs}$	Cub. $Ia\bar{3}$	a=9.365Å Z=16	$Mn_I(8b)$ $Mn_{II}(24d)$ O(48e)	$Mn_2^{[6]}[O_3\square]^{Cs}$ BIXBYITE	Wyckoff,2,2-4;Pov.,271;Str.Tab.,184, Am.Min.,1989,74,1325-1336;Acta Cryst.,1971,B27,821-828;LF,139..
BOWIEITE	$(Rh,Ir,Pt)_2S_3$	$(Rh,Ir,Pt)_2^{o}[S_3]^{n}$	Orth. Pbcn	a=8.462Å Z=4 b=5.985Å c=6.138Å	Rh(8d) $S_I(8d)$ $S_{II}(4c)$		Acta Cryst.,1967,23,832-840;Can. Min.,1984,22,543-552;Hölzel,29.
CLAUDETITE	As_2O_3		Mon. $P2_1/n$	a=7.99Å β=78.3° b=4.65Å Z=4 c=9.12Å	$As_{I-II}(4e)$ $O_{I-III}(4e)$		SR,41A,213-214;Pov.,309;Str.Tab., 182;Hölzel,63.
CORUNDUM	Al_2O_3	$Al_2^{o}[O_3]^{n}$	Trig. $R\bar{3}c$	a=4.7628Å a_R=5.128Å c=13.0032Å α=55°20' Z=6 Z_R=2	Al(12c) O(18e)	$Al_2[O_3]^{n}$ CORUNDUM	Wyckoff,2,68;Pov.,270;Acta Cryst., 1980,B36,228-230;LF,99.
ESKOLAITE	Cr_2O_3	$Cr_2^{o}[O_3]^{n}$	Trig. $R\bar{3}c$	a=4.954Å a_R=5.350Å c=13.584Å α=55°9' Z=6 Z_R=2	Cr(12c) O(18e)	$Al_2^{o}[O_3]^{n}$ CORUNDUM	Wyckoff,2,6-8;Am.Min.,1958,43,1098 -1106.
GUANAJUATI-TE	Bi_2Se_3	$Bi_2^{[3+4]}[Se_3]'$	Orth. Pbnm	a=11.35Å Z=4 b=11.48Å c=4.04Å		$Sb_2^{[3+4]}[S_3]'$ STIBNITE	Am.Min.,1950,35,337-364;Pov.,246, 128.
HEAZLEWOODI-TE	Ni_3S_2	$Ni_3^{t}[S_2]^{b}$	Trig. R32	a=5.74Å a_R=4.0718Å c=7.139Å α=89°25' Z=3 Z_R=1	Ni(3e) S(2c) (rhom.des.)		Str.Tab.,112;Am.Min.,1977,62,341-345;K/S,54-55;Pov.,221;SR,46A,103.
HEMATITE	Fe_2O_3 α-form	$Fe_2^{o}[O_3]^{n}$	Trig. $R\bar{3}c$	a=5.035Å a_R=5.4135Å c=13.72Å α=55°17' Z=6 Z_R=2	Fe(12c) O(18e)	$Al_2^{o}[O_3]^{n}$ CORUNDUM	Wyckoff,2,6-8;Am.Min.,1966,51,123-129;SR,31A,124;LF,99.
KARELIANITE	V_2O_3	$V_2^{o}[O_3]^{n}$	Trig. $R\bar{3}c$	a=5.105Å a_R=5.647Å c=14.449Å α=53°45' Z=6 Z_R=2		$Al_2^{o}[O_3]^{n}$ CORUNDUM	Wyckoff,2,6-8;Am.Min.,1963,48,33-41;Pov.,270;LF,99.
KASHINITE	$(Ir,Rh)_2S_3$	$(Ir,Rh)_2^{o}[S_3]^{b}$	Orth. Pbcn	a=8.450Å Z=4 b=6.001Å c=6.145Å			Am.Min.,1987,72,223(Abs.);Pearson, 1972,564-565.
LAPHAMITE	$As_2(Se,S)_3$	$2\infty[As_2^{[3n]}(Se,S)_3]$	Mon. $P2_1/n$	a=11.86Å β=90.17° b=9.756Å Z=4 c=4.265Å		$2\infty[As_2^{[3n]}S_3]$ ORPIMENT	Min.Mag.,1986,50,279-282;Wyckoff,2,26-27;LF,225.

Table 21

A_2B_3 (cont.)

NAME	CHEMICAL FORMULA	STRUCTURAL FORMULA	SPACE GROUP	UNIT CELL DIMENSIONS	EQUIVALENT POSITIONS	STRUCTURE TYPE	REFERENCES
MONTBRAYITE	$(Au,Sb)_2Te_3$		Tric. $P\bar{1}$	a=12.10Å α=104°30.5' b=13.46Å β=97°34.5' c=10.80Å γ=107°53.5' Z=12			Pov.,210;Str.Tab.,131;Hölzel,31.
MOSCHELLANDS-BERGITE	Ag_2Hg_3 γ-form	$[Ag_2,Hg_3]^b$	Cub. $Im\bar{3}m$	a=10.1Å Z=10			Am.Min.,1938,23,761-764; Pov.,195;RRW,420.
ORPIMENT	As_2S_3	$2\infty[As_2{}^{[3n]}S_3]$	Mon. $P2_1/n$	a=11.475Å β=90°41' b=9.577Å Z=4 c=4.256Å	$As_{I-II}(4e)$ $S_{I-III}(4e)$	$2\infty[As_2{}^{[3n]}S_3]$ ORPIMENT	Zeit.Krist.,1972,136,48-65; Wyckoff,2,26-27;Pov.,259-260; LF,225.
OTTEMANNITE	Sn_2S_3		Orth. Pnma	a=8.878Å Z=4 b=3.751Å c=14.020Å	$Sn_{I-II}(4c)$ $S_{I-III}(4c)$		Acta Cryst.,1982,B38,2022-2023;Acta Cryst.,1967,23,471-476;Pov.,248;Wyckoff,2,428-431.
PARAGUANAJUATI-TE	$Bi_2(Se,S)_3$		Trig. $R\bar{3}m$	a=4.125Å Z=3 c=28.56Å			RRW,460;Hölzel,29;Pov.,261.
PLUMBOPALLADINI-TE	Pb_2Pd_3	$(Pb_2,\square)^o[Pd_3]^h$	Hex.	a=4.470Å c=5.719Å Z=1		Defect deriv. $Ni^o[As]^h$ NICKELINE	Am.Min.,1971,56,1121(Abs.); Hölzel,7.
SENARMONTITE	Sb_2O_3	$\{g\}[Sb_4{}^{[3]}O_6]$	Cub. $Fd\bar{3}m$	a=11.1519Å Z=16	Sb(32e) O(48f)		Acta Cryst.,1975,B31,2016-2018; Pov.,297;SR,42A,233-234;Wyckoff,2,16-19.
STANNOPALLADINI-TE	Pd_3Sn_2		Hex. $P6_3/mmc$	a=4.382Å Z=1 c=5.658Å			Can.Min.,1981,19,599-605; Hölzel,7.
STIBNITE	Sb_2S_3	$Sb_2{}^{[3+4]}[S_3]$ #	Orth. Pbnm	a=11.299Å (25°C) b=11.310Å Z=4 c=3.8389Å	$Sb_{I-II}(4c)$ $S_{I-III}(4c)$	$Sb_2{}^{[3+4]}[S_3]$ # STIBNITE	Wyckoff,2,27-28;Min.Abs.,77-1483;Zeit.Krist.,1975,142,447-449;Pov.,246;LF,160.
TAZHERANITE	$(Zr,Ti,Ca)_4O_6$		Cub. $Fm\bar{3}m$ ...	a=5.108Å Z=1			Am.Min.,1970,55,318(Abs.); RRW,606;Hölzel,80.
TELLURANTIMONY	Sb_2Te_3	$Sb_2{}^o[Te_3]^{chh}$	Trig. $R\bar{3}m$	a=4.25Å a_R=10.426Å c=30.4Å α=23°31' Z=3 Z_R=1	$Te_I(3a)$ $Te_{II}(6c)$ Sb(6c)	$Bi_2{}^o[Te_3]^{chh}$ TELLUROBISMU-THITE	Wyckoff,2,29-31;Min.Abs.,88M/1826;Can.Min.,1973,12,55-60; LF,102.
TELLUROBISMUTHI-TE	Bi_2Te_3	$Bi_2{}^o[Te_3]^{chh}$	Trig. $R\bar{3}m$	a=4.3835Å a_R=10.473Å c=30.487Å α=24°10' Z=3 Z_R=1	$Te_I(3a)$ $Te_{II}(6c)$ Bi(6c)	$Bi_2{}^o[Te_3]^{chh}$ TELLUROBISMU-THITE	Wyckoff,2,29-31;Min.Abs.,88M/1826;LF,101.
TONGBAITE	Cr_3C_2		Orth. Pnma	a=5.5329Å Z=4 b=2.8290Å c=11.4719Å	$Cr_{I-III}(4c)$ $C_{I-II}(4c)$		SR,34A,56-57;Am.Min.,1985,70,218(Abs.);Bokii,363;Wyckoff,2,36-37.
UMANGITE	Cu_3Se_2	$(Cu_3\square)^t[Se_2]^c$ (Defect deriv. Digenite)	Tet. $P\bar{4}2_1m$	a=6.024Å c=4.2786Å Z=2	$Cu_I(2a)$ $Cu_{II}(4e)$ Se(4e)	$(Cu_3\square)^t[Se_2]^c$ UMANGITE	SR,42A,75-76;Pov.,232;Hölzel,13.
VALENTINITE	Sb_2O_3	$1\infty[Sb_2{}^{[3]}O_3]$	Orth. Pccn	a=4.911Å Z=4 b=12.464Å c=5.412Å	Sb(8e) $O_I(4c)$ $O_{II}(8e)$	$1\infty[Sb_2{}^{[3]}O_3]$ VALENTINITE	Acta Cryst.,1974,B30,458-461; Wyckoff,2,20-21;Pov.,300.

Table 22

A_2B_3(cont.)

NAME	CHEMICAL FORMULA	STRUCTURAL FORMULA	SPACE GROUP	UNIT CELL DIMENSIONS	EQUIVALENT POSITIONS	STRUCTURE TYPE	REFERENCES
WEISHANITE	$(Au,Ag)_3Hg_2$		Hex. $P6_3/mmc$	a=2.9265Å Z=2 c=4.8178Å			Am.Min.,1988,<u>73</u>,196(Abs.); Bokii,301-302;Hölzel,2.

Table 23

AB_4

NAME	CHEMICAL FORMULA	STRUCTURAL FORMULA	SPACE GROUP	UNIT CELL DIMENSIONS	EQUIVALENT POSITIONS	STRUCTURE TYPE	REFERENCES
DURANUSITE	As_4S	$^2_\infty[(As_4,S)^{[3n]}]^{-H\,c}$	Orth.	a=3.576Å Z=2 b=6.759Å c=10.074Å		Dist.deriv. $^2_\infty[As^{[3n]}]^{-H\,c}$ ARSENIC	Bull.Min.,1973,<u>96</u>,131-134; Hölzel,48.
PATRONITE	VS_4	$^1_\infty[V^{[8]}S_4]$	Mon. I 2/c	a=6.78Å β=100.8° b=10.42Å Z=8 c=12.11Å	V(8f) S_{I-IV}(8f)	$^1_\infty[V^{[8]}S_4]$ PATRÓNITE	SR,<u>29</u>,89-90;Pov.,245-246; Hölzel,46.
ROALDITE	$(Fe,Ni)_4N$	$N^o[(Fe,Ni)_4]^c$	Cub. $Pm\bar{3}m$	a=3.79Å Z=1	Fe_I(1a) N(1b) Fe_{II}(3c)	$N^o[(Fe,Ni)_4]^c$ ROALDITE	Am.Min.,1981,<u>66</u>,1100(Abs.); SR,<u>22</u>,155;Wyckoff,<u>2</u>,137-138;LF,103.

Table 24

A_3B_4

NAME	CHEMICAL FORMULA	STRUCTURAL FORMULA	SPACE GROUP	UNIT CELL DIMENSIONS	EQUIVALENT POSITIONS	STRUCTURE TYPE	REFERENCES
BORNHARDTI-TE	Co_3Se_4	$Co_2{}^{o}Co^{t}[Se_4]^{c}$	Cub. $Fd\bar{3}m$	a=10.20Å Z=8		$Al_2{}^{o}Mg^{t}[O_4]^{c}$ SPINEL	Am.Min.,1956,41,164-165(Abs.); Pov.,223;Wyckoff,3,75-78;LF,120.
BREZINAITE	Cr_3S_4		Mon. I 2/m	a=5.973Å β=91°9' b=3.432Å Z=2 c=11.361Å	$Cr_I(2a)$ $Cr_{II}(4i)$ $S_{I-II}(4i)$		Wyckoff,2,150-151;Am.Min.,1969, 54,1509-1518;Hölzel,27.
DIMORPHITE-I	As_4S_3 α-form	{g}$[As_4S_3]$	Orth. Pnma...	a=9.07Å Z=4 b=8.01Å c=10.30Å	$As_{I-II}(4c)$ $S_I(8d)$ $As_{III}(8d)$ $S_{II}(4c)$		Zeit.Krist.,1973,138,161-166; SR,35A,140-141;SR,38A,29.
DIMORPHITE-II	As_4S_3 β-form	{g}$[As_4S_3]$	Orth. Pnma...	a=11.24Å Z=4 b=9.90Å c=6.56Å	$As_{I-II}(4c)$ $S_I(8d)$ $As_{III}(8d)$ $S_{II}(4c)$		Zeit.Krist.,1973,138,161-166;SR, 39A,17-18.
GENKINITE	$(Pt,Pd)_4Sb_3$		Tet. P422	a=7.736Å Z=8 c=24.161Å			Can.Min.,1977,15,389-392;Hölzel, 11;Am.Min.,1979,64,654(Abs.).
GREIGITE	Fe_3S_4	$Fe_2{}^{o}Fe^{t}[S_4]^{c}$	Cub. $Fd\bar{3}m$	a=9.876Å Z=8	$Fe^{o}(16d)$ S(32e) $Fe^{t}(8a)$	$Al_2{}^{o}Mg^{t}[O_4]^{c}$ SPINEL	Am.Min.,1964,49,543-555; Hölzel,27;LF,120.
HAUSMANNITE	Mn_3O_4	$Mn_2{}^{o}Mn^{t}[O_4]^{c}$	Tet. $I\,4_1/amd$	a=5.765Å Z=4 c=9.442Å	$Mn^{o}(8d)$ O(16h) $Mn^{t}(4a)$	Dist.deriv. $Al_2{}^{o}Mg^{t}[O_4]^{c}$ SPINEL	SR,54A,141;Hölzel,61;LF,121.
IKUNOLITE	$Bi_4(S,Se)_3$	$Bi_4{}^{o}[(S,Se)_3]^{c/n}$	Trig. $R\bar{3}m$	a=4.15Å a_R=13.28Å c=39.19Å α=18°00' Z=3 Z_R=1	$(S,Se)_I(3a)$ $(S,Se)_{II}(6c)$ $Bi_{I-II}(6c)$		Am.Min.,1960,45,477-478(Abs.); SR,23,55;Pov.,261.
LAITAKARITE	$Bi_4(Se,S)_3$	$Bi_4{}^{o}[(Se,S)_3]^{c/n}$	Trig. $R\bar{3}m$	a=4.225Å a_R=13.53Å c=39.93Å α=17°58' Z=3 Z_R=1			Am.Min.,1962,47,806-807(Abs.); SR,23,55;Pov.,261;Can.Min., 1963,7,677-678.
LINNAEITE	Co_3S_4	$Co_2{}^{o}Co^{t}[S_4]^{c}$	Cub. $Fd\bar{3}m$	a=9.457Å Z=8	$Co^{o}(16d)$ S(32e) $Co^{t}(8a)$	$Al_2{}^{o}Mg^{t}[O_4]^{c}$ SPINEL	SR,11,288;Am.Min.,1964,49; 543-555;LF,120.
MAGHEMITE	$Fe_{2.67}O_4$	$(Fe_{1.67},\square_{0.3})^{o}Fe^{t}[O_4]^{c}$	Cub. $P2_13$	a=8.339Å Z=11		Defect deriv. $Al_2{}^{o}Mg^{t}[O_4]^{c}$ SPINEL	RRW,372;Hölzel,64;Am.Min., 1988,73,153160;Min.Abs., 89/0892:Pov.,186;Str.Tab.,186.
MAGNETITE	Fe_3O_4	$Fe_2{}^{o}Fe^{t}[O_4]^{c}$	Cub. $Fd\bar{3}m$	a=8.3969Å Z=8	$Fe^{o}(16d)$ O(32e) $Fe^{t}(8a)$	$Al_2{}^{o}Mg^{t}[O_4]^{c}$ SPINEL	Acta Cryst.,1984,C40,1491-1493;Wyckoff,3,75-79;LF,120.
MINIUM	Pb_3O_4	$Pb_2{}^{[3n]}\{1\infty\}[Pb^{o}O_4]$	Tet. $P4_2/mbc$	a=8.811Å Z=4 c=6.563Å	$Pb_I(4d)$ $O_I(8h)$ $Pb_{II}(8h)$ $O_{II}(8g)$	$Pb_2{}^{[3n]}\{1\infty\}[Pb^{o}O_4]$ MINIUM	SR,41A,212-213;Pov.,306-307; LF,204.
PARAMELA-CONITE	Cu_4O_3	$Cu_2{}^{sq}Cu_2{}^{[2]}[O_3]^{Qs}$	Tet. $I\,4_1/amd$	a=5.837Å Z=4 c=9.932Å	$Cu_I(8c)$ $O_I(8e)$ $Cu_{II}(8d)$ $O_{II}(4a)$		Am.Min.,1978,63,180185;SR,44A, 187;Acta Cryst.,1978,B34,22-26.
PILSENITE	Bi_4Te_3		Trig. $R\bar{3}m$	a=4.446Å Z=3 c=41.94Å			Am.Min.,1984,69,215(Abs.); Hölzel,30.
POLYDYMITE	Ni_3S_4	$Ni_2{}^{o}Ni^{t}[S_4]^{c}$	Cub. $Fd\bar{3}m$	a=9.48Å Z=8	$Ni^{o}(16d)$ S(32e) $Ni^{t}(8a)$	$Al_2{}^{o}Mg^{t}[O_4]^{c}$ SPINEL	Wyckoff,3,81;Pov.,223;LF,120.

Table 25

A_3B_4(cont.)

NAME	CHEMICAL FORMULA	STRUCTURAL FORMULA	SPACE GROUP	UNIT CELL DIMENSIONS	EQUIVALENT POSITIONS	STRUCTURE TYPE	REFERENCES
RUCKLIDGEITE	$(Bi,Pb)_3Te_4$	$(Bi,Pb)_3{}^{o}[Te_4]^{c/n}$	Trig. $R\bar{3}m$	a=4.422Å Z=3 c=41.49Å			Am.Min.,1978,63,599(Abs.); Hölzel,30.
TRÜSTEDTITE	Ni_3Se_4	$Ni_2{}^{o}Ni^{t}[Se_4]^{c}$	Cub. $Fd\bar{3}m$	a=9.94Å Z=8		$Al_2{}^{o}Mg^{t}[O_4]^{c}$ SPINEL	Am.Min.,1965,50,519-520 (Abs.);Pov.,223;LF,120.
TYRRELLITE	$(Cu,Co,Ni)_3Se_4$	$(Co,Ni)_2{}^{o}(Cu,Ni)^{t}[Se_4]^{c}$	Cub. $Fd\bar{3}m$	a=10.01Å Z=8		$Al_2{}^{o}Mg^{t}[O_4]^{c}$ SPINEL	Pov.,223;Hölzel,27;LF,120.
WILKMANITE	Ni_3Se_4		Mon. I 2/m	a=6.22Å β=90.53° b=3.63Å Z=2 c=10.52Å			Am.Min.,1965,50,519(Abs.); Pov.,225;Str.Tab.,128;RRW, 670-671.

Table 26

A_2B_5

NAME	CHEMICAL FORMULA	STRUCTURAL FORMULA	SPACE GROUP	UNIT CELL DIMENSIONS	EQUIVALENT POSITIONS	STRUCTURE TYPE	REFERENCES
KOUTEKITE	Cu_5As_2		Hex.	a=11.51Å Z=18 c=14.54Å			Am.Min.,1961,46,467(Abs.);Pov.,204; R/S,168-169;Hölzel,10.
ORCELITE	$Ni_{5-x}As_2$		Hex.	a=6.815Å Z=6 c=12.498Å			Am.Min.,1960,45,753-754;Pov.,204; Hölzel,16.
PERRYITE	$(Ni,Fe)_5(Si,P)_2$		Hex.				Min.Mag.,1968,36,850-854;Hölzel,4.
SHCHERBINAITE	V_2O_5	$V_2{}^{[5+1]}[O_5]^{c}$	Orth. Pmmn	a=11.519Å Z=2 b=3.564Å c=4.373Å	V(4f) O_{III}(2a) O_{I-II}(4f)		Wyckoff,2,184-187;Am.Min.,1973, 58,560(Abs.);Acta Cryst.,1986,42, 1467-1465;Bokii,440.
STIBIOPALLADINITE	Pd_5Sb_2		Hex. $P6_3/mmc$...	a=7.598Å Z=12 c=28.112Å			Am.Min.,1976,61,1249-1254;Hölzel,11.
TANTITE	Ta_2O_5		Tric. P1 ...	a=3.80Å α=90.92° b=3.79Å β=90.18° c=35.74Å γ=90° Z=6			Am.Min.,1984,69,1193(Abs.);Hölzel, 81.

AB_6

NAME	CHEMICAL FORMULA	STRUCTURAL FORMULA	SPACE GROUP	UNIT CELL DIMENSIONS	EQUIVALENT POSITIONS	STRUCTURE TYPE	REFERENCES
ALGODONITE	Cu_6As		Hex. $P6_3/mmc$	a=2.590Å Z=1 c=4.234Å			SR,16,24;Pov.,205;Hölzel,10;Can. Min.,1990,28,751-755.

Table 27

$A_mB_n(m\geq3,n\geq5)$

NAME	CHEMICAL FORMULA	STRUCTURAL FORMULA	SPACE GROUP	UNIT CELL DIMENSIONS	EQUIVALENT POSITIONS	STRUCTURE TYPE	REFERENCES
ALACRANITE	As_8S_9		Mon. P2/c	a=9.8.9Å β=101.84° b=9.73Å Z=2 c=9.13Å			Am.Min.,1988,73,189 (Abs.);Hölzel,suppl..
ANILITE	Cu_7S_4	$(Cu_7\square)^{t/tr}[S_4]^c$	Orth. Pnma	a=7.89Å Z=4 b=7.84Å c=11.01Å	$Cu_{I-III}(4c)$ $Cu_{IV-V}(8d)$ $S_{I-II}(4c)$ $S_{III}(8d)$	Dist.defect deriv. $Cu_2^{t/tr}[S]^c$ DIGENITE	Acta Cryst.,1970,B26,915-924;Am.Min.,1969.54,1256-1268;LF,109;Can.Min.,1985,23,61-76.
ARSENOPALLADINITE	$Pd_8(As,Sb)_3$		Tric. P1 ...	a=7.43 Å α=92° 53' b=13.95Å β=119° 30' c=7.35Å γ=87° 51' Z=6			Am.Min.,1979,64,658(Abs.); Hölzel,11.
ATHABASCAITE	Cu_5Se_4	$(Cu_5,\square_3)^{t/tr}[Se_4]^c$	Orth.	a=8.227Å Z=4 b=11.982Å c=6.441Å		Dist.defect deriv, $Cu_2^{t/tr}[S]^c$ DIGENITE	Can.Min.,1970,10,207-215; Pov.,232.
COBALT PENTLANDITE	$(Co,Ni,Fe)_9S_8$	$(Co,Ni,Fe)^o$ $(Co,Ni,Fe)_8{}^t[S_8]^c$	Cub. $Fm\bar{3}m$	a=9.923Å Z=4	$(Co,Ni.Fe)_I(4b)$ $(Co,Ni,Fe)_{II}(32f)$ $S_I(8c)$ $S_{II}(24e)$	$(Ni,Fe)^o(Ni,Fe)_8{}^t[S_8]^c$ PENTLANDITE	Can.Min.,1975,13,75-78; Pov.,226;LF,131.
DJURLEITE	$Cu_{31}S_{16}$	$(Cu_{31},\square)^{t/t}[S_{16}]^h$	Mon. $P2_1/n$	a=26.897Å β=90.13° b=15.745Å Z=8 c=13.565Å	$Cu_{I-LXII}(4e)$ $S_{I-XXXII}(4e)$	Defect deriv. $Cu_2^{tr/t}[S]^h$ CHALCOCITE	Zeit.Krist.,1979,150,299-320;Am.Min.,1962,47,1181-1184;Can.Min.,1985,23,61-76.
FERRIHYDRITE	$Fe_{4\text{-}5}(OH,O)_{12}$		Hex.	a=5.08Å Z=1 c=9.4Å			Am.Min.,1975,60,485-489; Hölzel,84.
GEERITE	Cu_8S_5	$(Cu_8,\square_2)^{t/tr}[S_5]^c$	Cub. $F\bar{4}3m$	a=5.410Å Z=4		Dist.defect deriv. $Cu_2^{t/tr}[S]^c$ DIGENITE	Can.Min.,1980,18,519-523; Can.Min.,1985,23,61-76; Hölzel,12:LF,109.
GEFFROYITE	$(Cu,Fe,Ag)_9$ $(Se,S)_8$	$(Cu,Fe,Ag)^o$ $(Cu,Fe,Ag)_8{}^t$ $[(Se,S)_8]^c$	Cub. $Fm\bar{3}m$	a=10.889Å Z=4		$(Ni,Fe)^o(Ni,Fe)_8{}^t[S_8]^c$ PENTLANDITE	Am.Min.,1982,67,1074-1075 (Abs.);Hölzel,16.
GODLEVSKITE	$(Ni,Fe)_9S_8$	$(Ni,Fe)_5{}^t(Ni,Fe)_4{}^{[4n]}$ $[S_8]^c$	Orth. C222	a=9.3359Å Z=4 b=11.2185Å c=9.4300Å	$Ni_I(4e)$ $Ni_{IV}(4k)$ $Ni_{II}(4b)$ $N_V(4h)$ $Ni_{III}(4i)$ ...		Acta Cryst.,1987,C43,2255-2257;Can.Min.,1988,26,283-291;Pov.,221.
HAXONITE	$(Fe,Ni)_{23}C_6$		Cub.	a=10.55Å Z=4			Am.Min.,1974,59,209(Abs.); Hölzel,4;Bokii,361.
ISOMERTIEITE	$Pd_{11}(Sb,As)_4$		Cub. $Fd\bar{3}m$	a=12.283Å Z=8			Min.Mag.,1974,39,528-543; Hölzel,11.
KEITHCONNITE	$Pd_{20}Te_7$		Trig. $R\bar{3}$ ...	a=11.45Å Z=3 c=11.40Å			Hölzel,suppl.;Encyclopedia Miner.Names,155.

Table 28

$A_mB_n(m\geq3,n\geq5)$(cont.)

NAME	CHEMICAL FORMULA	STRUCTURAL FORMULA	SPACE GROUP	UNIT CELL DIMENSIONS	EQUIVALENT POSITIONS	STRUCTURE TYPE	REFERENCES
KHARAELAKHITE	$(Cu,Pt,Pb,Fe,Ni)_9S_8$	$(Cu,Pt,Pb,Fe,Ni)^o(Cu,Pt,Pb,Fe,Ni)_8^t[S_8]^c$	Orth. Pmmm ...	a=9.713Å Z=4 b=8.333Å c=14.50Å		Dist.deriv. $(Ni,Fe)^o(Ni,Fe)_8^t[S_8]^c$ PENTLANDITE	Min.Zhurn.,1985,7,78-; Hölzel,16.
KOLYMITE	Cu_7Hg_6		Cub. $I\bar{4}3m$	a=9.418Å Z=4	Hg(24g) O_{I-II}(8c) O_{III}(12e)		ZVMO,1980,109,206-; Bokii,311-312;Hölzel,2.
MACKINAWITE	$(Fe,Ni)_9S_8$		Tet. $P4_2/nmm$	a=3.676Å Z=2 c=5.032Å			Am.Min.,1963,48,511-524; Hölzel,24;Pov.,220.
MAUCHERITE	$Ni_{11}As_8$		Tet. $P4_12_12$	a=6.8724Å Z=4 c=21.821Å	Ni_{I-V}(8b) As_{I-II}(4a) Ni_{VI}(4a) As_{III-V}(8b)		Am.Min.,1973,58,203-210; SR,39A,12;Hölzel,16.
MERTIEITE-I	$Pd_{11}(Sb,As)_4$		Hex.	a=15.04Å Z=18 c=22.41Å			Am.Min.,1973,58,1-10; Hölzel,suppl..
MERTIEITE-II	$Pd_8(Sb,As)_3$		Trig.	a=7.546Å Z=12 c=43.18Å			Can.Min.,1975,13,321-335; Hölzel,suppl.;K/S,158.
NOVÁKITE	$(Cu,Ag)_{21}As_{10}$		Mon. C2	a=16.269Å β=112°42' b=11.711Å Z=4 c=10.007Å			Hölzel,suppl.;Pov.,203.
OOSTERBOSCHITE	$(Pd,Cu)_7Se_5$		Orth.	a=10.42Å Z=8 b=10.60Å c=14.43Å			Bull.Min.,1970,93,476-481; Hözel,12;Pov.,221.
PALLADSEITE	$Pd_{17}Se_{15}$		Cub. Pm3m	a=10.635Å Z=2			Min.Mag.,1977,41,123; Hölzel,12;K/S,53-54.
PENTLANDITE	$(Ni,Fe)_9S_8$	$(Ni,Fe)^o(Ni,Fe)_8^t[S_8]^c$	Cub. $Fm\bar{3}m$	a=10.044Å Z=4	$(Ni,Fe)^o$(4b) S_I(8c) $(Ni,Fe)^t$(32f) S_{II}(24e)	$(Ni,Fe)^o(Ni,Fe)_8^t[S_8]^c$ PENTLANDITE	Can.Min.,1973,12,169-177; SR,42A,116-117;LF,131.
PYRRHOTITE-4C	Fe_7S_8	$(Fe_7,\square)^o[S_8]^h$	Mon. C2/c ...	a=11.90Å β=90°30' b=6.87Å Z=8 c=22.88Å		Defect deriv. $Ni^o[As]^h$ NICKELINE	Acta Cryst.,1969,B25,673-676;Econ.Geol.,1975,70,824-833;K/S,28;LF,86.
PYRRHOTITE-5C	Fe_9S_{10}	$(Fe_9,\square)^o[S_{10}]^h$	Hex.	a=6.888Å Z= c=28.670Å		Defect deriv. $Ni^o[As]^h$ NICKELINE	Econ.Geol.,1975,70,824-833;K/S,28;LF,86.
PYRRHOTITE-6C	$Fe_{11}S_{12}$	$(Fe_{11},\square)^o[S_{12}]^h$	Mon. Fd ...	a=6.8950Å β=90.00° b=11.9536Å Z=8 c=34.518Å		Defect deriv. $Ni^o[As]^h$ NICKELINE	Acta Cryst.,1975,B31,2759-2764;K/S,28;LF,86.
PYRRHOTITE-11C	$Fe_{10}S_{11}$	$(Fe_{10},\square)^o[S_{11}]^h$	Hex.	a=6.8969Å Z= c=63.222Å		Defect deriv. $Ni^o[As]^h$ NICKELINE	Econ.Geol.,1975,70,824-833;LF,86.
RICKARDITE	Cu_7Te_5	$[Cu_7Te_5]^{(2Q6)T}$	Tet. P4/nmm	a=3.98Å c=6.12Å Z=(0.4)			K/S,26-27;Am.Min.,1949,34,441-451;Hölzel,13;Pov.,216.

Table 29

$A_mB_n(m\geq3,n\geq5)$(cont.)

NAME	CHEMICAL FORMULA	STRUCTURAL FORMULA	SPACE GROUP	UNIT CELL DIMENSIONS	EQUIVALENT POSITIONS	STRUCTURE TYPE	REFERENCES
SMYTHITE	$(Fe,Ni)_9S_{11}$	$((Fe,Ni)_9\square_2)^o[S_{11}]^h$	Trig. $R\bar{3}m$	a=3.465Å a_R=11.67Å c=34.34Å α=17°06′ Z=3 Z_R=1		Defect deriv. $Ni^o[As]^h$ NICKELINE	Am.Min.,1970,55,1650-1658;Pov.,264;K/S,21;LF,86.
SPIONKOPITE	$Cu_{39}S_{28}$		Trig. $P\bar{3}m1$ ...	a=22.962Å c=41.429Å Z=18			Can.Min.,1980,18,511-518; Can.Min.,1985,23,61-76; Hölzel,12.
STILLWATERITE	Pd_8As_3		Hex. $P\bar{3}$ ...	a=7.399Å Z=3 c=10.311Å			Can.Min.,1975,13,321-335; K/S,158;Hölzel,11.
STÜTZITE	$Ag_{5-x}Te_3$		Hex. P6/mmm	a=13.38Å Z=7 c=8.45Å			Str.Tab.,111;Hölzel,14.
TELLUROPALLADI-NITE	Pd_9Te_4		Mon. $P2_1/c$	a=7.45Å β=91.9° b=13.94Å Z=4 c=8.82Å			Can.Min.,1979,17,589-594; Hölzel,11.
UZONITE	As_4S_5		Mon. $P2_1/m$	a=7.98Å β=100.14° b=8.10Å Z=2 c=7.09Å			Am.Min.,1986,71,1280(Abs.); Hölzel,46.
XIFENGITE	Fe_5Si_3		Hex. $P6_3/mmc$	a=6.759Å Z=2 c=4.720Å	Fe_I(4d) Si(6g) Fe_{II}(6g)		Am.Min.,1986,71,228(Abs.); Hölzel,4;SR,10,63.
YARROWITE	Cu_9S_8		Trig. $P\bar{3}m1$ ...	a=3.800Å Z=3 c=67.26 Å			Can.Min.,1985,23,61-76; Hölzel,12.

Table 30

ABC

NAME	CHEMICAL FORMULA	STRUCTURAL FORMULA	SPACE GROUP	UNIT CELL DIMENSIONS	EQUIVALENT POSITIONS	STRUCTURE TYPE	REFERENCES
AKAGANÉITE	FeO(OH,Cl) β-form	$Fe_8^{o}[(Cl,H_2O)$ $(O,OH)_{16}]^{chh}$	Tet. I 4/m	a=10.48Å c=3.023Å Z=8	Fe(8h) $(Cl,H_2O)(2b)$. $(O,OH)_{I-II}(8h)$	$Mn_8^{o}[(Ba,K)\ O_{16}]^{chh}$ HOLLANDITE	Acta Cryst.,1979,A35,197-200; Wyckoff,3,494-496;Min.Mag., 1960,32,545-557;LF,107.
ALLOCLASITE	(Co,Fe)AsS	$(Co,Fe)^{o}$ $\{g\}[AsS]^{c}$	Mon. $P2_1$	a=4.661Å β=90.2° b=5.602Å Z=2 c=3.411Å	(Co,Fe)(2a) As(2a) S(2a)	Dist.deriv. $Co^{o}\{g\}[AsS]^{c}$ COBALTITE	SR,43A,20;K/S,34;Hölzel,34.
ARSENOPYRITE	FeAsS	$Fe^{o}\{g\}[AsS^{«h»}]^{c}$ (Subs.deriv. Marcasite)	Mon. $B2_1/d$	a=9.451Å β=89.84° b=5.6492 Z=8 c=6.5456Å	Fe(8e) As(8e) S(8e)	$Fe^{o}\{g\}[AsS^{«h»}]^{c}$ ARSENOPYRITE	Zeit.Krist.,1987,179,335-346; Wyckoff,1,356-358;LF,170.
BISMOCLITE	BiOCl	$Bi^{[9]}$ $[[Cl][O]]^{(Q2,2Q1\ f)\ f}$	Tet. P4/nmm	a=3.819Å (26°C) c=7.369Å Z=2	Bi(2c) Cl(2a) O(2c)	$Pb^{[9]}[[F][Cl]]^{(Q2,2Q1\ f)\ f}$ MATLOCKITE	Wyckoff,1,294-296;Pov.,653-654;Hölzel,56;LF,157.
BÖHMITE	AlO(OH)	$Al^{o}[O(OH)]^{R31\ my}$ (relat.Lepidocro- cite)	Orth. Bbmm	a=12.227Å Z=4 b=3.700Å c=2.866Å	Al(4c) O(4c) (OH)(4c)		Wyckoff,1,292-294;Pov.,322-323;LF,146.
BRACEWELLITE	CrO(OH)	$Cr^{o}[O(OH)]^{n}$	Orth. Pbnm	a=4.492Å Z=4 b=9.860Å c=2.974Å		$Al^{o}[O(OH)]^{n}$ DIASPORE	Am.Min.,1977,62,593;Wyckoff, 1,290-291,Pov.,317;LF,96.
COBALTITE	CoAsS	$Co^{o}\{g\}[AsS]^{c}$ (Subs.deriv. Pyrite)	Orth. $Pca2_1$	a=5.592Å Z=4 b=5.587Å c=5.567Å	Co(4a) Ag(4a) S(4a)	$Co^{o}\{g\}[AsS]^{c}$ COBALTITE	Am.Min.,1982,67,1048-1057; Am.Min.,1965,50,1002-1014; Can.Min.,1986,24,27-33;LF,171.
COSTIBITE	CoSbS	$Co^{o}\{g\}[SbS^{«h»}]^{c}$	Orth. $Pn2_1m$	a=4.873Å Z=2 b=5.852Å c=3.608Å	Co(2a) Sb(2a) S(2a)	Dist.deriv. $Fe^{o}\{g\}[AsS^{«h»}]^{c}$ ARSENOPYRITE	SR,42A,14-16;Am.Min.,1970, 55,10-17;Pov.,243-244;K/S, 34-35.
DAUBRÉEITE	BiO(OH,Cl)	$Bi^{[9]}$ $[[Cl][OH]]^{(Q2,2Q1\ f)\ f}$	Tet. P4/nmm	a=3.85Å Z=2 c=7.40Å		$Pb^{[9]}[[F][Cl]]^{(Q2,2Q1\ f)\ f}$ MATLOCKITE	RRW,165;Pov.,653-654,734; Wyckoff,1,294;LF,157.
DIASPORE	AlO(OH) α-form	$Al^{o}[O(OH)]^{n}$	Orth. Pbnm	a=4.4007Å Z=4 b=9.4253Å c=2.8452Å	Al(4c) H(4c) $O_{I-II}(4c)$	$Al^{o}[O(OH)]^{n}$ DIASPORE	Min.Abs.,80-2872;Wyckoff,1, 290-291;Pov.,317;LF,96.
EUCAIRITE	CuAgSe	$Cu^{t}Ag^{t}[Se]^{c}$	Orth.	a=4.105Å Z=10 b=20.35Å c=6.31Å			RRW,198;Pov.,241;Hölzel,14; Ross,763;K/S,50.
FEITKNECHTITE	MnO(OH) β-form		Tet.	a= Z= c=4.62Å			Am.Min.,1965,50,1296-1354; Hölzel,82;Str.Tab.,216;Pov.,322.
FEROXYHYTE	FeO(OH) δ-form		Hex.	a=2.93Å Z=1 c=4.60Å			Am.Min.,1977,62,1057-1061; Hölzel,82.
GERSDORFFITE-Pa3	NiAsS	$Ni^{o}\{g\}[(As,S)_2]^{c}$	Cub. $Pa\bar{3}$	a=5.7053Å Z=4	Ni(4a) (As,S)(8c)	$Fe^{o}\{g\}[S_2]^{c}$ PYRITE	Can.Min.,1986,24,27-33;SR, 33A,359-361;LF,171.
GERSDORFFITE-P2₁3	NiAsS	$Ni^{o}\{g\}[[AsS]^{c}$	Cub. $P2_13$	a=5.576Å Z=4	Ni(4a) (As,S)(4a) (S,As)(4a)	$Ni^{o}\{g\}[SbS]^{c}$ ULLMANNITE	SR,41A,23;K/S,32;Hölzel,32; SR,32A,26-27;Pov.,244;Can. Min.,1986,24,27-33.

Table 31

ABC(cont.)

NAME	CHEMICAL FORMULA	STRUCTURAL FORMULA	SPACE GROUP	UNIT CELL DIMENSIONS	EQUIVALENT POSITIONS	STRUCTURE TYPE	REFERENCES
GERSDORFFITE-Pca2$_1$	NiAsS	Ni^{o}{g}[(As,S)$_2$]^{c} (Subs.deriv. Pyrite)	Orth. Pca2$_1$	a=5.6849Å Z=4 b= c=	Ni$_{I-IV}$(1a) (As,S)$_{I-VIII}$(1a)	Ni^{o}{g}[(As,S)$_2$]^{c} COBALTITE	Min.Mag.,1968,36,940-947;Am.Min., 1982,67,1048-1057,1058-1064;Can. Min.,1986,24,27-33,Hölzel,32.
GLAUCODOT	(Co,Fe)AsS	(Co,Fe)^{o} {g}[AsS$^{«h»}$]^{c}	Orth. Cmmm	a=6.64Å Z=24 b=28.39Å c=5.64Å		Dist.deriv. Fe^{o}{g}[AsS$^{«h»}$]^{c} ARSENOPYRITE	JCPDS,5-643;RRW,238-239; Pov.,245;Hölzel,34.
GOETHITE	FeO(OH) α-form	Fe^{o}[O(OH)]^{h}	Orth. Pbnm	a=4.64Å Z=4 b=10.0Å c=3.03Å	Fe(4c) O$_{I-II}$(4c) H(4c)	Al^{o}[O(OH)]^{h} DIASPORE	Wyckoff,1,290-291;SR,33A, 266-267;Pov.,317.
GRIMALDIITE	CrO(OH)	Cr^{o}[O(OH)]$^{(21s)c}$	Trig. R$\bar{3}$m	a=2.973Å c=13.392Å Z=3	(rhomb.des.) Cr(1a) (O,OH)(2c)	Co^{o}[O(OH)]$^{(21s)c}$ HETEROGENI-TE-3R	Am.Min.,1977,62,593(Abs.); SR,18,515-516;Pov.,322;LF,152.
GROUTITE	MnO(OH) α-form	Mn^{o}[O(OH)]^{h}	Orth. Pbnm	a=4.56Å Z=4 b=10.70Å c=2.85Å	Mn(4c) O$_{I-II}$(4c)	Al^{o}[O(OH)]^{h} DIASPORE	Wyckoff,1,290-292:Acta Cryst., 1968,B24,1233-1236;Pov.,317.
GUDMUNDITE	FeSbS	Fe^{o}{g}[SbS$^{«h»}$]^{c}	Mon. B2$_1$/d	a=10.04Å β=90° b=5.93Å Z=8 c=6.68Å	Fe(8e) Sb(8e) S(8e)	Fe^{o}{g}[AsS$^{«h»}$]^{c} ARSENOPYRI-TE	Wyckoff,1,356-357:JCPDS, 8-104;Pov.,245;Hölzel,34.
GUYANAITE	CrO(OH)	Cr^{o}[O(OH)]^{h}	Orth. Pnnm	a=4.857Å Z=2 b=4.295Å c=2.958Å		Dist.deriv. Mn^{o}[O(OH)]^{h} MANGANITE	Am.Min.,1977,62,593(Abs.);Pov., 317;Wells.1984,639;Pov.,317.
HETEROGENITE-2H	CoO(OH)	Co^{o}[O(OH)]$^{(21s)h}$	Hex. P6$_3$/mmc	a=2.85Å c=8.80Å Z=2	Co(2a) (O,OH)(4f)	Co^{o}[O(OH)]$^{(21s)h}$ HETEROGENI-TE-2H	Min.Mag.,1973,39,152-157; Hölzel,84.
HETEROGENITE-3R	CoO(OH)	Co^{o}[O(OH)]$^{(21s)c}$	Trig. R$\bar{3}$m	a=2.894Å a$_R$=4.676Å c=13.130Å α=35°28' Z=3 Z$_R$=1	(rh.desc.) Co(1a) (O,OH)(2c)	Co^{o}[O(OH)]$^{(21s)c}$ HETEROGENI-TE-3R	SR,18,515-516;Min.Mag.,1973, 39,152-157;LF,152.
HOLLINGWOR-THITE	(Rh,Pt,Pd)AsS	(Rh,Pt,Pd)^{o} {g}[(As,S)$_2$]^{c}	Cub. Pa$\bar{3}$	a=5.94Å Z=4		Fe^{o}{g}[S$_2$]^{c} PYRITE	Am.Min.,1965,50,1068-1074; Pov.,244;LF,171.
IRARSITE	(Ir,Ru,Rh,Pt) AsS	(Ir,Ru,Rh,Pt)^{o} {g}[(As,S)$_2$]^{c}	Cub. Pa$\bar{3}$	a=5.777Å Z=4		Fe^{o}{g}[S$_2$]^{c} PYRITE	Am.Min.,1967,52,1580(Abs.); Pov.,244;LF.171.
KITKAITE	NiTeSe	Ni^{o}[(Te,Se)$_2$]^{h}	Trig. P$\bar{3}$m1	a=3.716Å Z=1 c=5.126Å	Ni(1a) (Te,Se)(2d)	Ni^{o}[Te$_2$]^{h} MELONITE	Am.Min.,1965,50,581-586;Pov., 259;LF,98.
LAURIONITE	PbCl(OH)	Pb$^{[9]}$ [[Cl][OH]]$^{(Q2,2Q1 f) f}$	Orth. Pcmn	a=9.6987Å Z=4 b=4.0203Å c=7.1110Å	Pb(4c) Cl(4c) O(4c) H(4c)		Zeit.Krist.,1975,141,246-259;SR, 41A,13;Wells,1962,397-398;LF, 157;Pov.,654-655.
LAUTITE	CuAsS	Cu^{t}[AsS]^{c} (Subs.deriv. Sphalerite)	Orth. Pna2$_1$	a=11.35Å Z=4 b=5.456Å c=3.749Å	Cu(4a) As(4a) S(4a)		Acta Cryst.,1965,19,543-547;K/S,45;Pov.,256-257.
LEPIDOCROCITE	FeO(OH) γ-form	Fe^{o}[O(OH)]$^{R31\ my}$	Orth. Cmc2$_1$	a=3.08Å Z=4 b=12.50Å c=3.87Å	Fe(4a) O$_{I-II}$(4a) H(4a)		SR,44A,274-275;Pov.,322-323; LF,146.

Table 32

ABC(cont.)

NAME	CHEMICAL FORMULA	STRUCTURAL FORMULA	SPACE GROUP	UNIT CELL DIMENSIONS	EQUIVALENT POSITIONS	STRUCTURE TYPE	REFERENCES
MAJAKITE	PdNiAs		Hex. P¯62m	a=6.066Å Z=6 c=7.20Å			Am.Min.,1977,62,1260(Abs.); Hölzel,11.
MANGANITE	MnO(OH) γ-form	Mn^{o}[O(OH)]^{h} (Subs.deriv.Rutile)	Mon. B2$_1$/d	a=8.88Å β=90° b=5.25Å Z=8 c=5.71Å	Mn(8f) O$_{I-II}$(8f)	Mn^{o}[O(OH)]^{h} MANGANITE	Zeit.Krist.,1963,118,303-326; Pov.,318;LF,89.
MASLOVITE	PtBiTe	Pt^{o}{g}[(Bi,Te)$_2$]^{c}	Cub. Pa$\bar{3}$	a=6.687Å Z=4	Pt(4a) (Bi,Te)(8c)	Fe^{o}{g}[S$_2$]^{c} PYRITE	Am.Min.,1989,74,1168-1176; Hölzel,33.
MATLOCKITE	PbClF	Pb$^{[9]}$ [[F][Cl]]$^{(Q2,2Q1 f) f}$	Tet. P4/nmm	a=4.106Å Z=2 c=7.23Å	Pb(2c) F(2a) Cl(2c)	Pb$^{[9]}$[[F][Cl]]$^{(Q2,2Q1 f) f}$ MATLOCKITE	Wyckoff,1,295-296;Wells,1962, 390-391,674;LF,157;Pov.,644-5.
MICHENERITE	(Pd,Pt)BiTe	(Pd,Pt)^{o}{g}[BiTe]^{c} (Subs. deriv.Pyrite)	Cub, P2$_1$3	a=6.642Å Z=4	(Pd,Pt)(4a) Te(4a) Bi(4a)	Ni^{o}{g}[SbS]^{c} ULLMANNITE	Can.Min.,1973,12,61-65;Can. Min.,1986,24,27-33;Pov.,206; LF,171.
MONTROSEITE	(V,Fe)O(OH)	(V,Fe)^{o}[O(OH)]^{h}	Orth. Pbnm	a=4.54Å Z=4 b=9.97Å c=3.03Å	(V,Fe)(4c) O$_{I-II}$(4c)	Al^{o}[(O(OH)]^{h} DIASPORE	Wyckoff,1,290-292;SR,19, 364-365;Pov.,317;LF,96.
OSARSITE	(Os,Ru)AsS	(Os,Ru)^{o}{g}[AsS$^{«h»}$]^{c}	Mon. B2$_1$/c	a=5.933Å β=112°21' b=5.916Å Z=4 c=6.009Å		Fe^{o}{g}[AsS$^{«h»}$]^{c} ARSENOPYRITE	Am.Min.,1972,57,1029-1036; Hölzel,34.
PARACOSTIBITE	CoSbS	Co{g}[SbS]^{c}	Orth. Pbca	a=5.764Å Z=8 b=5.952Å c=11.635Å			Can.Min.,1970,10,232-246; Am.Min.,1971,56,631(Abs.): Pov.,245.
PARALAURIONITE	PbCl(OH)	Pb$^{[9]}$ [Cl(OH)]$^{(Q2,2Q1f)f}$	Mon. C2/m	a=10.79Å β=117°13' b=3.98Å Z=4 c=7.19Å			Pov.,654;RRW,461;Hölzel,56.
PETROVSKAITE	AuAg(S,Se)		Mon. P2m ...	a=4.943Å β=95°41' b=6.670Å Z=4 c=7.221Å			ZVMO,1984,113,602-607; Hölzel,14.
PLATARSITE	(Pt,Rh,Ru) AsS	(Pt,Rh,Ru)^{o} [g]{(As,S)$_2$]^{c}	Cub. Pa$\bar{3}$	a=5.788Å Z=4	(Pt,Rh,Ru)(4a) (As,S)(8c)	Fe^{o}{g}[S$_2$]^{c} PYRITE	Can.Min.,1979,17,117-123; SR,45A,27;Pov.,244.
STROMEYERITE	CuAgS	Cu^{tr}Ag^{tr}[S]^{c}	Orth. Cmcm	a=4.06Å Z=4 a=6.66Å c=7.99Å	Ag(4a) Cu(4c) S(4c)		SR.19,412-413;Pov.,241,126; Ross,763;Hölzel,13;K/S,50-51.
TOLOVKITE	IrSbS	Ir^{o}{g}[SbS]^{c} (Subs.deriv.Pyrite)	Orth. Pca2$_1$	a=6.025Å Z=4 b= c=	Ir(4a) Sb(4a) S(4a)	Co^{o}{g}[AsS]^{c} COBALTITE	Am.Min.,1989,74,1168-1176; Can.Min.,1986,24,27-33;Am. Min.,1982,67,1076 (Abs.).
ULLMANNITE	NiSbS	Ni^{o}{g}[SbS]^{c} (Subs.deriv.Pyrite)	Cub. P2$_1$3	a=5.911Å Z=4	Ni(4a) S(4a) Sb(4a)	Ni^{o}{g}[SbS]^{c} ULLMANNITE	Am.Min.,1980,65,154-156; SR,21,34;Pov.,244.
WILLYAMITE	(Co,Ni)SbS	(Co,Ni)^{o}{g}[SbS]^{c} (Subs.deriv.Pyrite)	Cub. P2$_1$3	a=5.9218Å Z=4	(Co,Ni)(4a) Sb(4a) S(4a)	Ni^{o}{g}[SbS]^{c} ULLMANNITE	Can.Min.,1986,24,27-33;SR, 46A,10;Pov.,244;Am.Min.,1980, 65,154-156.
ZAVARITSKITE	BiOF	Bi$^{[9]}$[[F][O]]$^{(Q2,2Q1 f) f}$	Tet. P4/nmm	a=3.75Å Z=2 c=6.23Å		Pb$^{[9]}$[[F][Cl]]$^{(Q2,2Q1 f) f}$ MATLOCKITE	Am.Min.,1963,48,210(Abs.); Pov.,653-654;LF,157.

Table 33

ABC_2

NAME	CHEMICAL FORMULA	STRUCTURAL FORMULA	SPACE GROUP	UNIT CELL DIMENSIONS	EQUIVALENT POSITIONS	STRUCTURE TYPE	REFERENCES
ARAMAYOITE	$Ag(Sb,Bi)S_2$		Tric. P1	a=7.76Å α=100°11' b=8.85Å β=90°43' c=8.23Å γ=103°50' Z=6	Ag_{I-VI}(1a) $(Sb,Bi)_{I-VI}$(1a) S_{I-XII}(1a)		Zeit.Krist.,1974,139,54-69; SR,40A,7;K/S,86-87;Pov., 238;Hölzel,23.
BOHDANOWICZI-TE	$AgBiSe_2$		Trig. P$\bar{3}$m1	a=4.183Å Z=3 c=19.561Å			Min.Mag.,1979,43,131-133; Hölzel,23.
CABRIITE	Pd_2CuSn	$[(Pd,Cu)_3Sn]^c$	Orth. Pmmm	a=7.88Å Z=4 b=7.88Å c=3.94Å			Can.Min.,1983,21,481-487; Hölzel,7.
CASWELLSILVE-RITE	$NaCrS_2$	$Na^o[Cr^{[2]}S_2]^{[3 1s) c}$ (=Delafossite)	Trig. R$\bar{3}$m	a=3.55Å Z=3 c=19.5Å			Am.Min.,1982,67,132-136; Hölzel,27;LF,152.
CHALCOPYRITE	$CuFeS_2$	$Cu^tFe^t[S_2]^c$	Tet. I $\bar{4}$2d	a=5.289Å Z=4 c=10.423Å	Cu(4a) S(8d) Fe(4b)	$Cu^tFe^t[S_2]^c$ CHALCOPYRITE	SR,39A,51-52;Pov.,227;LF,111.
CHALCOSTIBITE	$CuSbS_2$	$Cu^tSb^t[S_2]^n$	Orth. Pnam	a=6.026Å Z=4 b=14.510Å c=3.800Å			Can.Min.,1979,17,601-605; Pov.,255-256;K/S,61-62.
CREDNERITE	$CuMnO_2$	(related to Delafossite)	Mon. C2/m	a=5.530Å β=104.6° b=2.884Å Z=2 c=5.898Å	Mn(2a) Cu(2d) O(4i)		SR,22,334-335;Bull.Min., 1966,89,80-88;Pov.,313.
DELAFOSSITE	$CuFeO_2$	$Fe^o[Cu^{[2]}O_2]^{[3 1s) c}$	Trig. R$\bar{3}$m	a=3.03Å a_R=5.97Å c=17.13Å α=29°26' Z=3 Z_R=1	Cu(3a) Fe(3b) O(6c)		Pov.,313;SR,37A,262; Wyckoff,2,291-296;SB,1,272; SB,6,86.
DUTTONITE	$VO(OH)_2$		Mon. I 2/c	a=8.80Å β=90°40' b=3.95Å Z=4 c=5.96Å	V(4e) O_I(4e) O_{II}(8g) (in Imcm)		Acta Cryst.,1958,11,56-58; Am.Min.,1957,42,455-460; Pov.,321-322;SR,22,290.
EMPLECTITE	$CuBiS_2$	$Cu^tBi^{[3n]}[S_2]^n$	Orth. Pnma	a=6.1426Å Z=4 b=3.9189Å c=14.5282Å	Bi(4c) Cu(4c) S_I(4c) S_{II}(4c)		Zeit.Krist.,1975,141,387-402; K/S,61-62;SR,41A,34-35; Pov.,255-256.
ESKEBORNITE	$CuFeSe_2$	$(Cu,Fe)_2{}^t[Se_2]^c$	Tet. $P4_2mc$ ...	a=5.518Å Z=4 c=11.048Å		Dist.deriv. $Zn^t[S]^c$ SPHALERITE	Min.Abs.,89M/0899;Hölzel,19;Pov., 229;Am.Min.,1961,46,467(Abs.).
FERRONICKEL-PLATINUM	Pt_2FeNi		Tet.	a=3.871Å Z=1 c=3.635Å			Am.Min.,1984,69,1190-1191(Abs.); Hölzel,6;Can.Min.,1977,15,380-384.
GALLITE	$CuGaS_2$	$Cu^tGa^t[S_2]^c$	Tet. I $\bar{4}$2d	a=5.35Å c=10.48Å Z=4	Cu(4a) Ga(4b) S(8d)	$Cu^tFe^t[S_2]^c$ CHALCOPYRITE	Am.Min.,1959,44,906(Abs.);Wyckoff, 2,336-338;Str.Tab.,115;Pov.,227;LF, 111.
KOLARITE	$PbTeCl_2$		Orth. Pmmm ...	a=5.93Å Z=1 b=3.25Å c=3.89Å			Can.Min.,1985,23,501-506; Hölzel,47.
LORANDITE	$TlAsS_2$		Mon. $P2_1/a$	a=12.28 β=104°5'Å b=11.30 Z=8 c=6.10Å	Tl_{I-II}(4e) As_{I-II}(4e) S_{I-IV}(4e)		Zeit.Krist.,1973,138,147-160;SR, 39A,19-20;Pov.,257-258;K/S,76-77.
MATILDITE	$AgBiS_2$	$Ag^oBi^o[S_2]^c$ (sub.deriv.Halite)	Trig. P$\bar{3}$m1	a=8,12Å Z=12 c=19.06Å		$Ag^oBi^o[S_2]^c$ MATILDITE	Can.Min.,1969,9,655-662;Wyckoff,2, 334-335;Hölzel,23;Pov.,238.

Table 34

ABC_2(cont.)

NAME	CHEMICAL FORMULA	STRUCTURAL FORMULA	SPACE GROUP	UNIT CELL DIMENSIONS	EQUIVALENT POSITIONS	STRUCTURE TYPE	REFERENCES
MCCONNELLITE	$CuCrO_2$	$Cr^{o}[Cu^{[2]}O_2]^{[3 1s] c}$ (=Delafossite)	Trig. $R\bar{3}m$	a=2.983Å a_R=5.96Å c=17.160Å α=28°50' Z=3 Z_R=1			Am.Min.,1977,62,593-600;Pov.,313; Str.Tab.,175.
METABORITE	HBO_2 γ -form	$3\infty[B^{t}O_2H]$	Cub. $P\bar{4}3n$	a=8.886Å Z=24	H(24i) O_{I-II}(24i) B(24i)		Acta Cryst.,1963,16,380-384;Am.Min., 1965,50,261-262(Abs.);Pov.,464.
MIARGYRITE	$AgSbS_2$	$Ag^{[2/3]}Sb^{[3n]}[S_2]^{c}$	Mon. Cc	a=12.862Å β=98°38' b=4.411Å Z=8 c=13.220Å	Ag_{I-II}(4a) S_{I-IV}(4a) Sb_{I-II}(4a)		Acta Cryst.,1964,17,847-851; SR,29,21-22;Pov.,238;SR,6,83-84.
MUTHMANNITE	$AuAgTe_2$		?	?			Hölzel,31;Dokl.Earth Sci.,1985,280,159.
PUTORANITE	$Cu_{1.1}Fe_{1.2}S_2$	$Cu_{1.1}{}^{t}Fe_{1.2}{}^{t}[S_2]^{c}$	Cub.	a=5.30Å Z= ?			Am.Min.,1981,66,638(Abs.);Holzel,23; Garcia Orta,Ser.Geol.,1986,9,41-53.
RAGUINITE	$TlFeS_2$	$Tl^{t}Fe^{t}[S_2]^{c}$ (Subst.deriv. Sphalerte)	Orth.	a=12.40Å Z=8 b=10.44Å c=5.26Å		Dist.deriv. $Cu^{t}Fe^{t}[S_2]^{c}$ CHALCOPYRITE	Am.Min.,1969,54,1495(Abs.),1741 (Abs.);Pov.,227;Hölzel,19.
ROQUESITE	$CuInS_2$	$Cu^{t}In^{t}[S_2]^{c}$ (Subs. deriv.Sphalerite)	Tet. $I\bar{4}2d$	a=5.51Å Z=4 c=11.05Å	Cu(4a) S(8d) In(4a)	$Cu^{t}Fe^{t}[S_2]^{c}$ CHALCOPYRITE	Am.Min.,1963,48,1178-1179(Abs.); Wyckoff,2,336-338;Hölzel,18.
SCHAPBACHI-TE	$AgBiS_2$	$(Ag,Bi)_2{}^{o}[S_2]^{c}$	Cub.	a=5.65Å Z=2		$Na^{o}[Cl]^{c}$ HALITE	Str.Tab.,125:Hölzel,suppl.;Acta Cryst., 1959,12,46-54;Am.Min.,1960,45,591-8.
SMITHITE	$AgAsS_2$	$Ag^{[3]}As^{[2]}[S_2]^{c}$	Mon. C2/c	a=15.19Å β=101°12' b=7.78Å Z=24 c=17.23Å	Ag_{I-II}(8f) As_{I-III}(8f) Ag_{III-IV}(4e) S_{I-VI}(8f)		SR,29,25-26;Str.Tab.,141; Pov.,238.
TEALLITE	$PbSnS_2$	$(Pb,Sn)_2{}^{o}[S_2]^{c}$	Orth. Pmcn	a=4.05Å Z=2 b=4.29Å c=11.35Å		Dist.deriv. $Na^{o}[Cl]^{c}$ HALITE	Pov.,261-264;Hölzel,24;Str.Tab.,125.
TERLINGUAITE	Hg_2OCl	$1\infty[Hg_2OCl]$	Mon. C2/c	a=19.515Å β=143.81° b=5.915Å Z=8 c=9.478Å	Hg_{I}(4c) Hg_{III}(8f) Hg_{II}(4e) Cl(8f) O(8f)		SR,33A,221;SR,20,242-243;Pov.,652.
TRECHMANNITE	$AgAsS_2$	$(Ag_{18},As_{18},\square_3)^{o}[S_{36},\square_3]^{c}$	Trig. $R\bar{3}$	a=13.98Å a_R=8.62Å c=9.12Å α=108°17' Z=18 Z_R=6		Dist.defect deriv. $Na^{o}[Cl]^{c}$ HALITE	Zeit.Krist.,1969,129,163-177;K/S,86-88;Pov.238;Am.Min.,1968,53,1208-1214.
TULAMEENITE	Pt_2FeCu		Tet.	a=3.891Å Z=2 c=3.577Å			Can.Min.,1973,12,21-25;RRW,632; Hölzel,6.
VOLYNSKITE	$AgBiTe_2$		Trig. $R\bar{3}m$	a=4.27Å Z=3 c=29.1Å			Pov.,214;Am.Min.,1966,51,531 (Abs.);Hölzel,23;RRW,656..
WEISSBERGITE	$TlSbS_2$		Tric. P1	a=11.8Å α=109.9° b=6.4Å β=81.8° c=6.1Å γ =105.4° Z=4			Am.Min.,1978,63,720-724;K/S,76; Hölzel,39.
ZHARCHIKHITE	$AlF(OH)_2$		Mon. $P2_1/a$	a=5.164Å β=116.244° b=7.843Å Z=4 c=5.179Å			Am.Min.,1989,74,504(Abs.);Hölzel, 54;Min.Abs.,89M/0939.

Table 35

ABC_3

NAME	CHEMICAL FORMULA	STRUCTURAL FORMULA	SPACE GROUP	UNIT CELL DIMENSIONS	EQUIVALENT POSITIONS	STRUCTURE TYPE	REFERENCES
ALAMOSITE	$PbSiO_3$	$Pb^{[3/4]}\{1\infty\}[Si^tO_3]$	Mon. P2/n	a=11.23Å β=113°15' b=7.08Å Z=12 c=12.26Å	$Pb_{I-III}(4g)$ $Si_{I-III}(4g)$ $O_I(2c)$ $O_{II}(2f)$ $O_{III-X}(4g)$		Zeit.Krist.,1968,<u>126</u>,98-111; Pov.,412;SR,<u>33A</u>,454-455.
ARAGONITE	$CaCO_3$	$Ca^{[6/9]}\{g\}[C^{tr}O_3]$	Orth. Pmcn	a=4.9614Å Z=4 b=7.9671Å c=5.7404Å	Ca(4c) C(4c) $O_I(4c)$ $O_{II}(8d)$	$Ca^{[6/9]}\{g\}[C^{tr}O_3]$ ARAGONITE	Am.Min.,1971,<u>56</u>,758-767; Wyckoff,<u>2</u>,364-365;Pov.,610; SR,<u>37A</u>,280;LF,175.
BALYAKINITE	$CuTeO_3$		Orth. Pmcn	a=7.60Å Z=8 b=5.83Å c=12.69Å			Am.Min.,1981,<u>66</u>,436-439; Hölzel,92.
CALCITE	$CaCO_3$	$Ca^{ap}\{g\}[C^{tr}O_3^{«h»}]^c$	Trig. $R\bar{3}c$	a=4.9896Å a_R=6.361Å c=17.0610Å α=46°6' Z=6 Z_R=2	Ca(6b) C(6a) O(18e)	$Ca^{ap}\{g\}[C^{tr}O_3^{«h»}]^c$ CALCITE	Zeit.Krist.,1981,<u>156</u>,233-243; Wyckoff,<u>2</u>,360-362;Pov.,609; LF,172.
CERUSSITE	$PbCO_3$	$Pb^{[6/9]}\{g\}[C^{tr}O_3]$	Orth. Pmcn	a=5.1800Å Z=4 b=8.492Å c=6.134Å	Pb(4c) C(4c) $O_I(4c)$ $O_{II}(8d)$	$Ca^{[6/9]}\{g\}[C^{tr}O_3]$ ARAGONITE	Zeit.Krist.,1974,<u>139</u>,215-222; Pov.,610;LF,175.
CHLOROCALCITE	$KCaCl_3$		Orth. Pnma	a=7.551Å Z=4 b=10.442Å c=7.251Å			RRW,127;Hölzel,52.
CLINOENSTATITE	$MgSiO_3$	$Mg^{[6]}Mg^o$ $\{1\infty\}[Si_2{}^tO_6]$	Mon. $P2_1/c$	a=9.620Å β=108°20' b=8.825Å Z=8 c=5.188Å	$Mg_{I-II}(4e)$ $Si_{I-II}(4e)$ $O_{I-VI}(4e)$	Dist.deriv. $Ca^{[8]}Mg^o\{1\infty\}[Si_2{}^tO_6]$ DIOPSIDE	SR,<u>24</u>,465-466;Pov.,733, 408;Str.Tab.,411;SR,<u>43A</u>, 309;LF,205.
CLINOFERROSILITE	$(Fe,Mg)SiO_3$	$(Fe,Mg)^{[8]}(Fe,Mg)^o$ $\{1\infty\}[Si_2{}^tO_6]$	Mon. $P2_1/c$	a=9.72Å β=108°27' b=9.09Å Z=8 c=5.24Å		Dist.deriv. $Ca^{[8]}Mg^o\{1\infty\}[Si_2{}^tO_6]$ DIOPSIDE	Str.Tab.,411;Min.Mag., 1965,<u>34</u>,66-70.
ECANDREWSITE	(Zn,Fe,Mn) TiO_3	$(Zn,Fe,Mn)^oTi^o[O_3]^h$ (Subs.deriv. Corundum)	Trig. $R\bar{3}$	a=5.090Å a_R=5.50Å c=14.036Å α=55°7' Z=6 Z_R=2		$Fe^oTi^o[O_3]^h$ ILMENITE	Min.Mag.,1988,<u>52</u>,237-240; Hölzel,64.
ENSTATITE	$MgSiO_3$	$Mg^{[6]}Mg^o$ $\{1\infty\}[Si_2{}^tO_6]$	Orth. Pbca	a=18.235Å Z=16 b=8.818Å c=5.179Å	$Mg^{[6]}(8c)$ $Mg^o(8c)$ $Si_{I-II}(8c)$ $O_{I-VI}(8c)$	$Mg^{[6]}Mg^o\{1\infty\}[Si_2{}^tO_6]$ ENSTATITE	Zeit.Krist.,1986,<u>176</u>,159- 175;LF,207;Wyckoff,<u>4</u>,301- 303.
FAIRBANKITE	$PbTeO_3$		Tric. $P\bar{1}$	a=7.81Å α=117°12' b=7.11Å β=93°47' c=6.96Å γ=93°24' Z=4			Min.Mag.,1979,<u>43</u>,453- 457;Hölzel,92.
GASPÉITE	(Ni,Mg,Fe) CO_3	$(Ni,Mg,Fe)^{ap}$ $\{g\}[C^{tr}O_3^{«h»}]^c$	Trig. $R\bar{3}c$	a=4.621Å a_R=5.65Å c=14.93Å α=48°18' Z=6 Z_R=2	(Ni,Mg,Fe)(6b) C(6a) O(18e)	$Ca^{ap}\{g\}[C^{tr}O_3^{«h»}]^c$ CALCITE	Am.Min.,1966,<u>51</u>,677-684; Acta Cryst.,1986,<u>C42</u>,4-5; Str.Tab.,236-237;LF,172.
GEIKIELITE	$MgTiO_3$	$Mg^oTi^o[O_3]^h$ (Subs.deriv. Corundum)	Trig. $R\bar{3}$	a=5.10Å a_R=5.55Å c=14.12Å α=54°39' Z=6 Z_R=2		$Fe^oTi^o[O_3]^h$ ILMENITE	Str.Tab.,185;RRW,232-233; Pov.,738;
GETCHELLITE	$SbAsS_3$	$2\infty[Sb^{[3n]}As^{[3n]}S_3{}^{[2]}]$	Mon. $P2_1/a$	a=11.8568Å β=116.365° b=9.0152Å Z=8 c=10.1938Å	$(As,Sb)_{I-II}(4e)$ $(Sb,As)_{I-II}(4e)$ $S_{I-VI}(4e)$	$\{2\infty\}[Sb^{[3n]}As^{[3n]}S_3{}^{[2]}]$ GETCHELLITE	Acta Cryst.,1973,<u>B29</u>,2536- 2541;Am.Min.,1965,<u>50</u>,1817- 1826;Pov.,259-260;K/S,27.

Table 36

ABC_3(cont.)

NAME	CHEMICAL FORMULA	STRUCTURAL FORMULA	SPACE GROUP	UNIT CELL DIMENSIONS	EQUIVALENT POSITIONS	STRUCTURE TYPE	REFERENCES
ILMENITE	$FeTiO_3$	$Fe^{o}Ti^{o}[O_3]^{h}$ (Subs.deriv. Corundum)	Trig. $R\bar{3}$	a=5.082Å a_R=5.538Å c=14.026Å α=54.41° Z=6 Z_R=2	Fe(6c) Ti(6c) O(18f)	$Fe^{o}Ti^{o}[O_3]^{h}$ ILMENITE	Wyckoff,2,420-421;Acta Cryst.,1991,B47,561-580; RRW,298;LF,100.
KANOITE	$(Mn,Mg)SiO_3$	$(Mn,Mg)_2^{[6]}$ $\{1\infty\}[Si_2^{t}O_6]$ (relat.Diopside)	Mon. $P2_1/c$	a=9.739Å β=108.56° b=8.939Å Z=8 c=5.260Å			Am.Min.,1978,63,598(Abs.); Hölzel,212.
KIMROBINSONITE	$Ta(OH)_3(O,CO_3)$	Ta^{o} $[(O,CO_3)(OH)_3]^{c}$	Cub.	a=3.812Å Z=1			Can.Min.,1985,23,573-576; Am.Min.,1987,72,1024(Abs.)
LATRAPPITE	$(Ca,Na)(Nb,Ti)O_3$	$(Nb,Ti)^{o}$ $[(Ca,Na)O_3]^{c}$	Orth. Pcmn	a=5.448Å Z=4 b=7.777Å c=5.553Å		$Ti^{o}[CaO_3]^{c}$ PEROVSKITE	Can.Min.,1963,7,683-697; RRW,344;Pov.,295;Str.Tab., 188;LF,103.
LOPARITE-(Ce)	(Ce,Na,Ca) $(Ti,Nb)O_3$	$(Ti,Nb)^{o}$ $[(Ce,Na,Ca)O_3]^{c}$	Cub. $Pm\bar{3}m$	a=3.86Å Z=1		$Ti^{o}[CaO_3]^{c}$ PEROVSKITE (ideal)	Pov.,295;Str.Tab.,188; LF,103.
LUESHITE	$NaNbO_3$	$Nb^{o}[NaO_3]^{c}$	Orth. $P222_1$	a=5.51Å Z=8 b=5.53Å c=15.50Å		Dist.deriv. $Ti^{o}[CaO_3]^{c}$ (ideal) PEROVSKITE	Am.Min.,1961,46,1004(Abs.); Pov.,295;Hölzel,64.
MACEDONITE	$PbTiO_3$	$Ti^{o}[PbO_3]^{c}$	Tet. P4/mmm	a=3.889Å c=4.209Å Z=1		Dist.deriv. $Ti^{o}[CaO_3]^{c}$ (ideal) PEROVSKITE	Am.Min.,1971,56,387-394; Pov.,295;Hölzel,65.
MAGNESITE	$MgCO_3$	$Mg^{ap}\{g\}[C^{tr}O_3{}^{«h»}]^{c}$	Trig. $R\bar{3}c$	a=4.6328Å a_R=5.61Å c=15.0129Å α=48°11' Z=6 Z_R=2	Mg(6b) O(18e) C(6a)	$Ca^{ap}\{g\}[C^{tr}O_3{}^{«h»}]^{c}$ CALCITE	Zeit.Krist.,1981,156,233-243;Pov.,609;Hölzel,97.
MELANOSTIBITE	$Mn(Sb,Fe)O_3$	$Mn^{o}(Sb,Fe)^{o}[O_3]^{h}$ (Subs.deriv. Corundum)	Trig. $R\bar{3}$	a=5.226Å a_R=5.648Å c=14.325Å α=55°8' Z=6 Z_R=2	Mn(6c) (Sb,Fe)(6c) O(18f)	$Fe^{o}Ti^{o}[O_3]^{h}$ ILMENITE	Am.Min.,1968,53,1104-1109; Pov.,279;Str.Tab.,185;LF, 100.
MOLYBDOMENITE	$PbSeO_3$		Mon. $P2_1/m$	a=4.552Å β=106.40° b=5.525Å Z=2 c=6.633Å	Pb(2e) O_I(2e) Se(2e) O_{II}(4f)		SR,43A,285-286;Can.Min., 1965,8,149-159;Pov.,564; SR,38A,346;Str.Tab.,228.
NEIGHBORITE	$NaMgF_3$	$Mg^{o}[NaF_3]^{c}$	Orth. Pcmn	a=5.363Å Z=4 b=7.676Å c=5.503Å		$Ti^{o}[CaO_3]^{c}$ PEROVSKITE	Am.Min.,1961,46,379-393; Pov.,662-663;Str.Tab.,163.
NITER	KNO_3	$K^{[6/9]}\{g\}[N^{tr}O_3]$	Orth. Pcmn	a=5.43Å Z=4 b=9.19Å c=6.46Å	K(4c) N(4c) O_I(4c) O_{II}(8d)	$Ca^{[6/9]}\{g\}[C^{tr}O_3]$ ARAGONITE	Str.Tab.,233;Pov.,632; Wyckoff,2,364-365.
NITRAMMITE	NH_4NO_3	$(NH_4)^{[8+4]}$ $\{g\}[N^{tr}O_3]$	Orth. Pmmn	a=5.757Å Z=2 b=5.451Å c=4.935Å	(NH_4)(2b) N(2a) O_I(2a) O_{II}(4f)		Wyckoff,2,370-372;Acta Cryst.,1972,B28,1357-1361; Pov.,633;Min.Abs.,75-176.
NITRATINE	$NaNO_3$	$Na^{ap}\{g\}[N^{tr}O_3{}^{«h»}]^{c}$	Trig. $R\bar{3}c$	a=5.0708Å a_R=6.3247Å c=16.818Å α=47°16' Z=6 Z_R=2	Na(6b) N(6a) O(18e)	$Ca^{ap}\{g\}[C^{tr}O_3{}^{«h»}]^{c}$ CALCITE	Wyckoff,2,359-363;Hölzel, 96;Acta Cryst.,1967,23,455-460;Pov.,632;LF,172-173.

Table 37

ABC_3(cont.)

NAME	CHEMICAL FORMULA	STRUCTURAL FORMULA	SPACE GROUP	UNIT CELL DIMENSIONS	EQUIVALENT POSITIONS	STRUCTURE TYPE	REFERENCES
OTAVITE	$CdCO_3$	$Cd^{ap}\{g\}$ $[C^{tr}O_3^{«h»}]^c$	Trig. $R\bar{3}c$	a=4.9204Å a_R=6.1306Å c=16.298Å α=47°19' Z=6 Z=2	K(6b) O(18e) C(6a)	$Ca^{ap}\{g\}[C^{tr}O_3^{«h»}]^c$ CALCITE	Wyckoff,2,359-363;Pov.,609;Str.Tab.,236;LF,172-173.
PEROVSKITE	$CaTiO_3$	$Ti^o[CaO_3]^c$	Orth. Pcmn	a=5.3829Å Z=4 b=7.6453Å c=5.4458Å	Ti(4b) O_{II}(8d) Ca(4c) O_I(4c)	$Ti^o[CaO_3]^c$ PEROVSKITE	Acta Cryst.,1983,C39,1323-1325;Pov.,295;Wyckoff,2,410;LF,103.
PIGEONITE	(Mg,Fe,Ca) SiO_3	$(Ca,Fe)^{[7]}(Mg,Fe)^o$ $\{1\infty\}[Si_2{}^tO_6]$	Mon. $P2_1/c$	a=9.706Å β=108.59° b=8.950Å Z=8 c=5.246Å	(Ca,Fe)(4e) Si_{I-II}(4e) O_{I-VI}(4e) (Mg,Fe)(4e)	Dist.deriv. $Ca^{[8]}Mg^o\{1\infty\}[Si_2{}^tO_6]$ DIOPSIDE	Am.,Min.,1970,55,1195-1209;Wyckoff,4,298-299;LF,205.
PLUMBOTELLURITE	$PbTeO_3$ α-form		Orth.	a=8.423Å Z=12 b=13.739Å c=9.199Å			Am.Min.,1982,67,1075(Abs.);Hölzel,92.
POYARKOVITE	Hg_3OCl		Mon. C2/m…	a=18.82Å β=112°24' b=9.02Å Z=24 c=16.79Å			Am.Min.,1982,67,860(Abs.);Hölzel,55.
PYROPHANITE	$MnTiO_3$	$Mn^oTi^o[O_3]^h$	Trig. $R\bar{3}$	a=5.137Å a_R=5.62Å c=14.29Å α=54°16' Z=6 Z_R=2		$Fe^oTi^o[O_3]^h$ ILMENITE	RRW,501;Pov.,279;Hölzel,64;Str.Tab.,185;LF,100.
PYROXFERROITE	(Fe,Mn,Ca) SiO_3	$(Fe,Mn,Ca)_7{}^o$ $\{1\infty\}[Si_7{}^tO_{21}]$	Tric. P1…	a=6.62Å α=114.4° b=7.54Å β=82.7° c=17.35Å γ=94.5° Z=14		$(Fe,Mn,Ca)_7{}^o$ $\{1\infty\}[Si_7{}^tO_{21}]$ PYROXFERROITE	Am.Min.,1970,55,2137(Abs.);Pov.,412;Liebau,78-81,172-173;Hölzel,223;RRW,502.
PYROXMANGITE	$MnSiO_3$	$Mn_7{}^o\{1\infty\}[Si_7{}^tO_{21}]$	Tric. P1	a=6.671Å α=113.7° b=7.557Å β=84.0° c=17.45Å γ=94.3° Z=14		$Mn_7{}^o\{1\infty\}[Si_7{}^tO_{21}]$ PYROXMANGITE	SR,23,476-477;Acta Cryst.,1959,12,177-181;Pov.,412;Hölzel,223;Str.Tab.,427.
RHODOCHROSITE	$MnCO_3$	$Mn^{ap}[\{g\}[C^{tr}O_3^{«h»}]^c$	Trig. $R\bar{3}c$	a=4.7682Å a_R=5.91Å c=15.6354Å α=47°43' Z=6 Z_R=2	Mn(6b) C(6a) O(18e)	$Ca^{ap}\{g\}[C^{tr}O_3^{«h»}]^c$ CALCITE	Zeit.Krist.,1981,156,233-243;Pov.,609,756.
RHODONITE	(Mn,Fe,Mg,Ca) SiO_3	$(Mn,Fe,Mg,Ca)_5{}^{[7]}$ $\{1\infty\}[Si_5{}^tO_{15}]$	Tric. P'1	a=7.6816Å α=92.355° b=11.8180Å β=93.948° c=6.7073Å γ=105.665° Z=10	$(Mn.Fe.Mg,Ca)_{I-V}$ (2i) Si_{I-V}(2i) O_{I-XV}(2i)	$(Mn,Fe,Mg,Ca)_5{}^{[7]}$ $\{1\infty\}[Si_5{}^tO_{15}]$ RHODONITE	Wyckoff,4,319-320;Pov.,411;Am.Min.,1978,63,1137-1142;SR,28,258-259;SR,23,476-477;LF,215.
SCOTLANDITE	$PbSO_3$		Mon. $P2_1/m$…	a=4.542Å β=106.22° b=5.333Å Z=2 c=6.413Å			Min.Mag.,1984,48,283-288;Min.Abs.,86M/2937,309;Hölzel,92.
SIDERITE	$FeCO_3$	$Fe^{ap}\{g\}[C^{tr}O_3^{«h»}]^c$	Trig. $R\bar{3}c$	a=4.6916Å a_R=5.91Å c=15.3796Å α=47°43' Z=6 Z_R=2	Fe(6b) C(6a) O(18e)	$Ca^{ap}\{g\}[C^{tr}O_3^{«h»}]^c$ CALCITE	Zeit.Krist.,1981,156,233-243;Pov.,758,609;LF,172-173.
SMITHSONITE	$ZnCO_3$	$Zn^{ap}\{g\}[C^{tr}O_3^{«h»}]^c$	Trig. $R\bar{3}c$	a=4.6526Å a_R=5.68Å c=15.0257Å α=48°19' Z=6 Z_R=2	Zn(6b) C(6a) O(18e)	$Ca^{ap}\{g\}[C^{tr}O_3^{«h»}]^c$ CALCITE	Zeit.Krist.,1981,156,233-241;Pov.,609;LF,172-173.

Table 38

ABC_3(cont.)

NAME	CHEMICAL FORMULA	STRUCTURAL FORMULA	SPACE GROUP	UNIT CELL DIMENSIONS	EQUIVALENT POSITIONS	STRUCTURE TYPE	REFERENCES
SPHAEROCOBALTITE	$CoCO_3$	$Co^{6p}\{g\}[C^{tr}O_3^{\,eh_3}]^c$	Trig. $R\bar{3}c$	a=4.6618Å a_R=5.67Å c=14.963Å α=48°33' Z=6 Z_R=2	Co(6b) C(6a) O(18e)	$Ca^{6p}\{g\}[C^{tr}O_3^{\,eh_3}]^c$ CALCITE	Acta Cryst.,1986,C42,4-5; Pov.,609;Str.Tab.,236.
STRONTIANITE	$SrCO_3$	$Sr^{[6/9]}\{g\}[C^{tr}O_3]$	Orth. Pnma	a=6.020Å Z=4 b=5.093Å c=8.376Å	Sr(4c) C(4c) O_I(4c) O_{II}(8d)	$Ca^{[6/9]}\{g\}[C^{tr}O_3]$ ARAGONITE	SR,35A,295-296;Pov.,610; Str.Tab.,239:LF,175.
TAUSONITE	$SrTiO_3$	$Ti^o[SrO_3]^c$	Cub. $Pm\bar{3}m$	a=3.9048Å Z=1	Sr(1a) O(3c) Ti(1b)	$Ti^o[CaO_3]^c$ PEROVSKITE (ideal)	Am.Min.,1985,70,218(Abs.); Hölzel,65;LF,103.
VATERITE	$CaCO_3$	$Ca^{[6]}\{g\}[C^{tr}O_3]^h$	Hex. $P6_3/mmc$,,,	a=7.15Å c=16.94Å Z=12	Ca_{I-II}(2a) C(12j) Ca_{III}(12i) O_I(12j) O_{II}(24i) (part.occ.)	$Ca^{[6]}\{g\}[C^{tr}O_3]^h$ VATERITE	Zeit.Krist.,1969,128,183-212; Hölzel,98;Wyckoff,2,367-368; SR,34A,348-349.
WITHERITE	$BaCO_3$	$Ba^{[6/9]}\{g\}[C^{tr}O_3]$	Orth. Pmcn	a=5.3126Å Z=4 b=8.8958Å c=6.4284Å	Ba(4c) C(4c) O_I(4c) O_{II}(8d)	$Ca^{[6/9]}\{g\}[C^{tr}O_3]$ ARAGONITE	Am.Min.,1971,56,756-767; Pov.,610;Str.Tab.,239;LF, 175.
WOLLASTONITE-1T	$CaSiO_3$	$Ca^o\{1\infty\}[Si^tO_3]$	Tric. $P\bar{1}$	a=7.94Å α=90°2' Z=6 b=7.32Å β=95°22' c=7.07Å γ=103°26'		$Ca^o\{1\infty\}[Si^tO_3]$ WOLLASTONITE-1T	Str.Tab.,422;Am.Min.,1978, 63,274-288;Hölzel,218;Pov., 410-411.
WOLLASTONITE-2M	$CaSiO_3$	$Ca^o\{1\infty\}[Si^tO_3]$	Mon. $P2_1/a$	a=15.409Å β=95°30' b=7.322Å Z=12 c=7.063Å	Ca_{I-III}(4e) Si_{I-III}(4e) O_{I-IX}(4e)	$Ca^o\{1\infty\}[Si^tO_3]$ WOLLASTONITE-2M	Zeit.Krist.,1984,168,93-98; Wyckoff,4,265-266;LF,212; Hölzel,219;Str.Tab.,423.
WOLLASTONITE-3T	$CaSiO_3$	$Ca^o\{1\infty\}[Si^tO_3]$	Tric.	a=23.2Å α=90.0° b=7.30Å β=95.5° c=7.06Å γ=94.6° Z=		$Ca^o\{1\infty\}(SiO_3]$ WOLLASTONITE-3T	Am.Min.,1983,68,156-163.
WOLLASTONITE-4T	$CaSiO_3$	$Ca^o\{1\infty\}(Si^tO_3]$	Tric.	a=31.2Å α=90.0° b=7.30Å β=95.5° c=7.06Å γ=96.8° Z=		$Ca^o\{1\infty\}[Si^tO_3]$ WOLLASTONITE-4T	Am.Min.,1983,68,156-163.
WOLLASTONITE-5T	$CaSiO_3$	$Ca^o\{1\infty\}[Si^tO_3]$	Tric.	a=38.6Å α=90.0° b=7.30Å β=95.5° c=7.06Å γ=98.0° Z=		$Ca^o\{1\infty\}[Si^tO_3]$ WOLLASTONITE-5T	Am.Min.,1983,68,156-163.
WOLLASTONITE-7T	$CaSiO_3$	$Ca^o\{1\infty\}[Si^tO_3]$	Tric.	a=54.3Å α=90.0° b=7.30Å β=95.5° c=7.08Å γ=92.1° Z=42		$Ca^o\{1\infty\}[Si^tO_3]$ WOLLASTONITE-7T	Am.Min.,1979,74,658 (Abs.);Hölzel,219.

Table 39

ABC_4

NAME	CHEMICAL FORMULA	STRUCTURAL FORMULA	SPACE GROUP	UNIT CELL DIMENSIONS	EQUIVALENT POSITIONS	STRUCTURE TYPE	REFERENCES
AGUILARITE	Ag_4SeS	$3\infty[Ag_4^{[2]}(SeS)]$	Orth.	a=4.33Å Z= b=7.09Å c=7.76Å		Dist.subs.deriv. $3\infty\ [Cu_2^{[2]}S]$ CUPRITE	Can.Min.,1974,12,365-369; Pov.,725,233;Hölzel,14;LF,272.
ALUMOTANTITE	$AlTaO_4$		Orth.	a=4.90Å Z= b=11.58Å c=5.66Å			Am.Min.,1982,67,413(Abs.); Hölzel,79.
ANGLESITE	$PbSO_4$	$Pb^{[12]}\{g\}[S^tO_4]$	Orth. Pbnm	a=6.959Å Z=4 b=8.482Å c=5.398Å	Pb(4c) S(4c) $O_{I-II}(4c)$ $O_{III}(8d)$	$Ba^{[12]}\{g\}[S^tO_4]$ BARITE	Am.Min.,1978,63,506-510; Wyckoff,3,45-47;Pov.,581; LF,312.
ANHYDRITE	$CaSO_4$	$Ca^{[8]}S^t[O_4]^*$	Orth. Bbmn	a=6.238Å Z=4 b=6.991Å c=6.996Å	Ca(4c) S(4c) $O_I(8f)$ $O_{II}(8g)$	$Ca^{[8]}S^t[O_4]^*$ ANHYDRITE	Wyckoff,3,18-19;Pov.,581; LF,151.
AVOGADRITE	$(K,Cs)BF_4$	$(K,Cs)^{[12]}\{g\}[B^tF_4]$	Orth. Pnma	a=8.6588Å Z=4 b=5.4800Å c=7.0299Å	(K,Cs)(4c) B(4c) $F_{I-II}(4c)$ $F_{III}(8d)$	$Ba^{[12]}\{g\}[S^tO_4]$ BARITE	Acta Cryst.,1969,B25,2161-2162;Wyckoff,3,50;Pov.,665.
BARITE	$BaSO_4$	$Ba^{[12]}\{g\}[S^tO_4]$	Orth. Pbnm	a=7.157Å Z=4 b=8.884Å c=5.457Å	Ba(4c) S(4c) $O_{I-II}(4c)$ $O_{II}(8d)$	$Ba^{[12]}\{g\}[S^tO_4]$ BARITE	Am.Min.,1978,63,506-510; Wyckoff,3,45-47;Pov.,581; LF,312.
BEHIERITE	$(Ta,Nb)BO_4$	$(Ta,Nb)^{do}B^t[O_4]^*$	Tet. $I\ 4_1/amd$	a=6.206Å Z=4 c=5.472Å	(Ta,Nb)(4a) B(4b) O(16h)	$Zr^{do}Si^t[O_4]^*$ ZIRCON	Am.Min.,1962,47,414(Abs.);LF,149;Pov.,465;Wyckoff,3,15-16.
BERLINITE	$AlPO_4$	$3\infty[Al^tP^tO_4]$	Trig. $P3_121$	a=4.9429Å Z=3 c=10.9476Å	Al(3a) P(3b) $O_{I-II}(6c)$	$3\infty[Al^tP^tO_4]$ BERLINITE	SR,31A,183;Pov.,529;SR,29,367;Acta Cryst.,1964,17,82-85.
BETA-FERGUSONITE-(Ce)	$(Ce,La,Nd)NbO_4$	$(Ce,La,Nd)^{do}Nb^t[O_4]^*$	Mon.	a=5.170Å β=75°20' b=11.372Å Z=4 c=5.497Å (heat.700°C)			Am.Min.,1977,62,397(Abs.); Hölzel,77;Pov.,274.
BETA-FERGUSONITE-(Nd)	$(Nd,Ce)NbO_4$	$(Nd,Ce)^{do}Nb^t[O_4]^*$	Mon.	a=5.07Å β=93.0° b=5.62Å Z=2 c=5.41Å (heated)			Am.Min.,1984,69,406(Abs.); Hölzel,77;Pov.,274.
BETA-FERGUSONITE-(Y)	$YNbO_4$	$Y^{do}Nb^t[O_4]^*$	Mon. C2/c	a=7.645Å β=138.42° b=10.999Å Z=4 c=5.317Å	Y(4e) Nb(4e) $O_{I-II}(8f)$	Dist.deriv. $Ca^{do}W^t[O_4]^*$ SCHEELITE	Am.Min.,1961,46,1516-1517; SR,46A,251.
BEZSMERTNOVITE	$(Au,Ag)_4Cu(Te,Pb)$		Orth.	a=4.036Å Z= b=4.025Å c=4.061Å			Am.Min.,1981,66,878(Abs.); Hölzel,1.
BISMUTOTANTALITE	$Bi(Ta,Nb)O_4$	$Bi^o(Ta,Nb)^o[O_4]^h$	Orth. Pcmn...	a=4.957Å Z=4 b=11.76Å c=5.633Å		Dist.deriv. $(Fe,Mn)^oW^o[O_4]^h$ WOLFRAMITE	Am.Min.,1963,48,1348-1356; RRW,74-75;Pov.,274.
CELESTINE	$SrSO_4$	$Sr^{[12]}\{g\}[S^tO_4]$	Orth. Pnma	a=8.377Å Z=4 b=5.350Å c=6.873Å	Sr(4c) S(4c) $O_{I-II}(4c)$ $O_{III}(8d)$	$Ba^{[12]}\{g\}[S^tO_4]$ BARITE	SR,30A,365-366;Am.Min.,1978,63,506-510;Pov.,731,581;LF,312.
CERVELLEITE	Ag_4TeS		Cub. P23...	a=14.03Å Z=24			Euro.J.Min.,1989,1,371-380; Hölzel,14.

Table 40

ABC_4(cont.)

NAME	CHEMICAL FORMULA	STRUCTURAL FORMULA	SPACE GROUP	UNIT CELL DIMENSIONS	EQUIVALENT POSITIONS	STRUCTURE TYPE	REFERENCES
CHALCOCYANITE	$CuSO_4$	$Cu^{o}\{g\}[S^{t}O_4]$	Orth. Pnma	a=8.39Å Z=4 b=6.69Å c=4.83Å	Cu(4a) S(4c) O_{I-II}(4c) O_{III}(8d)		Acta Cryst.,1958,<u>11</u>,361-364; Hölzel,120;Pov.,580-581.
CHERALITE	(Th,Ca,Ce) $(P,Si)O_4$	$(Th,Ca,Ce)^{[9]}$ $(P,Si)^{t}[O_4]^{*}$	Mon. $P2_1/n$	a=6.751Å β=103°53' b=6.962Å Z=4 c=6.468Å		$(Ce,La,Nd,Th)^{[9]}$ $P^{t}[O_4]^{*}$ MONAZITE	Min.Mag.,1980,<u>43</u>,885-888; RRW,123;Pov.,732;Str.Tab., 313.
CHERNOVITE-(Y)	$YAsO_4$	$Y^{do}As^{t}[O_4]^{*}$	Tet. $I\,4_1/amd$	a=7.039Å Z=4 c=6.272Å		$Zr^{do}Si^{t}[O_4]^{*}$ ZIRCON	Am.Min.,1968,<u>53</u>,1777(Abs.); Pov.,508-509;LF,149.
CHROMATITE	$CaCrO_4$	$Ca^{do}Cr^{t}[O_4]^{*}$	Tet. $I\,4_1/amd$	a=7.26Å Z=4 c=6.26Å		$Zr^{do}Si^{t}[O_4]^{*}$ ZIRCON	Am.Min.,1964,<u>49</u>,439(Abs.); Pov.,575;RRW,129;LF.149.
CLINOBISVANITE	$BiVO_4$	$Bi^{do}V^{t}[O_4]^{*}$	Mon. I 2/a	a=5.186Å β=90°26' b=11.708Å Z=4 c=5.100Å			Min.Mag.,1974,<u>39</u>,847-849; Am.Min.,1963,<u>48</u>,1348-1356; Hölzel,147.
CROCOITE	$PbCrO_4$	$Pb^{[9]}Cr^{t}[O_4]^{*}$	Mon. $P2_1/n$	a=7.127Å β=102.43° b=7.438Å Z=4 c=6.799Å	Pb(4e) Cr(4e) O_{I-IV}(4e)	$(Ce,La,Nd,Th)^{[9]}$ $P^{t}[O4]^{*}$ MONAZITE	Zeit.Krist.,1986,<u>176</u>,75-83; Pov.,275;LF,321.
DREYERITE	$BiVO_4$	$Bi^{do}V^{t}[O_4]^{*}$	Tet. $I\,4_1/amd$	a=7.303Å Z=4 c=6.458Å		$Zr^{do}Si^{t}[O_4]^{*}$ ZIRCON	Am.Min.,1982,<u>67</u>,622(Abs.); Hölzel,146.
FERBERITE	$FeWO_4$	$Fe^{o}W^{o}[O_4]^{n}$	Mon. P2/c	a=4.72Å β=90° b=5.70Å Z=2 c=4.96Å	Fe(2f) W(2e) O_{I-II}(4g)	$(Fe,Mn)^{o}W^{o}[O_4]^{n}$ WOLFRAMITE	Zeit.Krist.,1968,<u>127</u>,61-72; Pov.,737,302;LF,92.
FERGUSONITE-(Y)	$YNbO_4$	$Y^{do}Nb^{t}[O_4]^{*}$	Tet. $I\,4_1/a$	a=5.16Å Z=4 c=10.89Å	Nb(4a) Y(4b) O(16f)	$Ca^{do}W^{t}[O_4]^{*}$ SCHEELITE	Str.Tab.,209;Pov.,274;SR,<u>23</u>, 388-389.
FERRUCCITE	$NaBF_4$	$Na^{[8]}B^{t}[F_4]^{*}$	Orth. Cmcm	a=6.8368Å Z=4 b=6.2619Å c=6.7916Å	Na(4c) B(4c) F_I(8f) F_{II}(8g)	$Ca^{do}S^{t}[O_4]^{*}$ ANHYDRITE	Acta Cryst.,1968,<u>B24</u>,1703-1704;SR,<u>33A</u>,181;Pov.,665.
FORMANITE-(Y)	$YTaO_4$	$Y^{do}Ta^{t}[O_4]^{*}$	Tet. $I\,4_1/a$	a=7.76Å Z=4 c=11.32Å (synthetic)		$Ca^{do}W^{t}[O_4]^{*}$ SCHEELITE	RRW,219;Str.Tab.,209:Pov., 274.
GASPARITE-(Ce)	(Ce,La,Nd) AsO_4		Mon. $P2_1/n$	a=6.937Å β=104.69° b=7.137Å Z=4 c=6.738Å			Am.Min.,1988,<u>73</u>,1494-1495 (Abs.);Hölzel,46.
HAFNON	$HfSiO_4$	$Hf^{do}Si^{t}[O_4]^{*}$	Tet. $I\,4_1/amd$	a=6.5725Å Z=4 c=5.9632Å	Hf(4a) Si(4b) O(16h)	$Zr^{do}Si^{t}[O_4]^{*}$ ZIRCON	Am.Min.,1982,<u>67</u>,804-808; Hölzel,187.
HASHEMITE	$Ba(Cr,S)O_4$	$Ba^{[12]}$ $\{g\}[(Cr,S)^{t}O_4]$	Orth. Pnma	a=9.1028Å Z=4 b=5.5276Å c=7.3314Å	Ba(4c) (Cr,S)(4c) O_{I-II}(4c) O_{III}(8d)	$Ba^{[12]}\{g\}[S^{t}O_4]$ BARITE	Am.Min.,1986,<u>71</u>,1217-1220; Acta Cryst.,1987,<u>C43</u>,1467-1469.
HETEROSITE	$(Fe,Mn)PO_4$	$(Fe,Mn)^{o}P^{t}[O_4]^{n}$ (relat. Forsterite)	Orth. Pmnb	a=5.83Å Z=4 b=9.79Å c=4.769Å	(Fe,Mn)(4c) P(4c) O_{I-II}(4c) O_{III}(8d)	$(Fe,Mn)^{o}P^{t}[O_4]^{n}$ HETEROSITE	Am.Min.,1972,<u>57</u>,45-51;Pov., 535-536;Hölzel,142; Angelova & Petrova,1996.
HÜBNERITE	$MnWO_4$	$Mn^{o}W^{o}[O_4]^{n}$	Mon. P2/c	a=4.2838Å β=91.18° b=5.7504Å Z=2 c=4.9901Å	Mn(2f) W(2e) O_{I-II}(4g)	$(Fe,Mn)^{o}W^{o}[O_4]^{n}$ WOLFRAMITE	Zeit.Krist.,1976,<u>144</u>,238-258; Pov.,302;LF,92.

Table 41

ABC_4(cont.)

NAME	CHEMICAL FORMULA	STRUCTURAL FORMULA	SPACE GROUP	UNIT CELL DIMENSIONS	EQUIVALENT POSITIONS	STRUCTURE TYPE	REFERENCES
HUTTONITE	$ThSiO_4$	$Th^{[9]}Si^{t}[O_4]^{*}$	Mon. $P2_1/n$	a=6.784Å β=104.92° b=6.974Å Z=4 c=6.500Å	Th(4e) Si(4e) O_{I-IV}(4e)	$(Ce,La,Nd,Th)^{[9]}P^{t}[O_4]^{*}$ MONAZITE	Acta Cryst.,1978,B34,1074-1078;Pov.,381;LF,321.
KOSTOVITE	$CuAuTe_4$		Orth. Pma2	a=16.50Å Z=4 b=8.84Å c=4.42Å			JCPDS,35-521;Pov.,213.
MONAZITE-(Ce)	(Ce,La,Nd,Th) PO_4	$(Ce,La,Nd,Th)^{[9]}P^{t}[O_4]^{*}$	Mon. $P2_1/n$	a=6.76Å β=103°38' b=7.00Å Z=4 c=6.44Å	(Ce,La,Nd,Th)(4e) P(4e) O_{I-IV}(4e)	$(Ce,La,Nd,Th)^{[9]}P^{t}[O_4]^{*}$ MONAZITE	Wyckoff,3,33-35;Pov.,536; LF,321.
MONAZITE-(La)	$(La,Ce,Nd)PO_4$	$(La,Ce,Nd)^{[9]}P^{t}[O_4]^{*}$	Mon. $P2_1/n$	a=6.84Å β=103.85° b=7.07Å Z=4 c=6.45Å		$(Ce,La,Nd,Th)^{[9]}P^{t}[O_4]^{*}$ MONAZITE	JCPDS,35-731.
MONAZITE-(Nd)	$(Nd,La,Ce)PO_4$	$(Nd,La,Ce)^{[9]}P^{t}[O_4]^{*}$	Mon. $P2_1/n$	a=6.745Å β=103.65° b=6.694Å Z=4 c=6.435Å		$(Ce,La,Nd,Th)^{[9]}P^{t}[O_4]^{*}$ MONAZITE	Am,Min,1988,73,1495(Abs.); Hölzel,146.
POWELLITE	$CaMoO_4$	$Ca^{do}Mo^{t}[O_4]^{*}$	Tet. $I\,4_1/a$	a=5.24Å Z=4 c=11.46Å		$Ca^{do}W^{t}[O_4]^{*}$ SCHEELITE	Str.Tab.,301;Pov.,569;LF,150.
PUCHERITE	$BiVO_4$	$1\infty[Bi^{do}V^{t}O_4]$	Orth. Pnca	a=5.328Å Z=4 b=5.052Å c=12.003Å	Bi(4c) V(4c) O_{I-II}(8d)	$1\infty[Bi^{do}V^{t}O_4]$ PUCHERITE	Zeit.Krist.,1984,169,289-294;Pov.,498-499.
PURPURITE	$(Mn,Fe)PO_4$	$(Mn,Fe)^{o}P^{t}[O_4]^{h}$ (relat. Forsterite)	Orth. Pmnb	a=5.83Å Z=4 b=9.70Å c=4.77Å		$(Fe,Mn)^{o}P^{t}[O_4]^{h}$ HETEROSITE	JCPDS,17-202;RRW,497; Pov.,755,535;Str.Tab.,308; Angelova & Petrova,1996.
RASPITE	$PbWO_4$	$Pb^{[7]}\{1\infty\}[W^{o}O_4]$	Mon. $P2_1/a$	a=13.555Å β=107.63° b=4.976Å Z=4 c=5.561Å	Pb(4e) W(4e) O_{I-IV}(4e)		Acta Cryst.,1977,B33,162-164;Pov.,570;Str.Tab.,204.
ROOSEVELTITE	$BiAsO_4$ α-form	$Bi^{[9]}As^{t}[O_4]^{*}$	Mon. $P2_1/n$	a=6.879Å β=104.84° b=7.159Å Z=4 c=6.732Å	Bi(4e) As(4e) O_{I-IV}(4e)	$(Ce,La,Nd,Th)^{[9]}P^{t}[O_4]^{*}$ MONAZITE	Acta Cryst.,1982,B38,1559-1561;Pov.,508-509.
SANMARTINITE	$(Zn,Fe)WO_4$	$(Zn,Fe)^{o}W^{o}[O_4]^{h}$	Mon. P2/c	a=4.72Å β≅90° b=5.75Å Z=2 c=4.97Å		$(Fe,Mn)^{o}W^{o}[O_4]^{h}$ WOLFRAMITE	Str.Tab.,204;Pov.,302; Hölzel,75.
SCHEELITE	$CaWO_4$	$Ca^{do}W^{t}[O_4]^{*}$	Tet. $I\,4_1/a$	a=5.243Å Z=4 c=11.376Å	Ca(4b) W(4a) O(16f)	$Ca^{do}W^{t}[O_4]^{*}$ SCHEELITE	SR,29,331-332;Pov.,569;LF, 150.
STIBIOCOLUMBITE	$SbNbO_4$	$Sb^{o}Nb^{[6]}[O_4]^{h}$	Orth. $Pc2_1/n$	a=4.929Å Z=4 b=11.79Å c=5.559Å		Dist.deriv. $(Fe,Mn)^{o}W^{o}[O_4]^{h}$ WOLFRAMITE	Am.Min.,1963,48,1348-1356; RRW,581;Pov.,274-275;Str. Tab.,210;LF,92.
STIBIOTANTALITE	$SbTaO_4$	$Sb^{o}Ta^{o}[O_4]^{h}$	Orth. $Pc2_1/n$	a=4.911Å Z=4 b=11.81Å c=5.535Å		Dist.deriv. $(Fe,Mn)^{o}W^{o}[O_4]^{h}$ WOLFRAMITE	Am.Min.,1963,48,1348-1356; RRW,581;Pov.,274-275;Str. Tab.,210;LF,92.
STOLZITE	$PbWO_4$ β-form	$Pb^{do}W^{t}[O_4]^{*}$	Tet. $I\,4_1/a$	a=5.45Å Z=4 c=12.03Å	Pb(4b) W(4a) O(16f)	$Ca^{do}W^{t}[O_4]^{*}$ SCHEELITE	Str.Tab.,301;Wyckoff,3, 19-23;Pov.,569;LF,150.

Table 42

ABC_4(cont.)

NAME	CHEMICAL FORMULA	STRUCTURAL FORMULA	SPACE GROUP	UNIT CELL DIMENSIONS	EQUIVALENT POSITIONS	STRUCTURE TYPE	REFERENCES
SYLVANITE	$AgAuTe_4$	$Ag^oAu^o[Te_4]^n$ (Subs.deriv. Melonite)	Mon. P2/c	a=8.95Å β=145.35° b=4.478Å Z=2 c=14.62Å	Au(2a) Ag(2e) Te_{I-II}(4g)		Min.Abs.,85M/2406,252;Pov., 213;SR,8,72-73;K/S,22.
THORITE	$ThSiO_4$	$Th^{do}Si^t[O_4]^*$	Tet. I 4_1/amd	a=7.1328Å Z=4 c=6.3188Å	Th(4a) O(16h) Si(4b)	$Zr^{do}Si^t[O_4]^*$ ZIRCON	Acta Cryst.,1978,B34,1074-1079;Pov.,381;LF,149.
WAKEFIELDITE-(Ce)	$(Ce,Pb)VO_4$	$(Ce,Pb)^{do}V^t[O_4]^*$	Tet. I 4_1/amd	a=7.35Å Z=4 c=6.56Å		$Zr^{do}Si^t[O_4]^*$ ZIRCON	Bull.Min.,1986,109,305;Hölzel, 146.
WAKEFIELDITE-(Y)	YVO_4	$Y^{do}V^t[O_4]^*$	Tet. I 4_1/amd	a=7.105Å Z=4 c=6.29Å		$Zr^{do}Si^t[O_4]^*$ ZIRCON	Am.Min.,1971,56,395-410; Hölzel,146;Pov.,493;LF,149.
WOLFRAMITE	$(Fe,Mn)WO_4$	$(Fe,Mn)^oW^o[O_4]^n$	Mon. P2/c	a=4.750Å β=90°10' b=5.720Å Z=2 c=4.970Å	W(2e) (Fe,Mn)(2f) O_{I-II}(4g)	$(Fe,Mn)^oW^o[O_4]^n$ WOLFRAMITE	Am.Min.,1971,56,489-498; SR,37A,256;Pov.,302;LF,92.
WULFENITE	$PbMoO_4$	$Pb^{do}Mo^t[O_4]^*$	Tet. I 4_1/a	a=5.4312Å c=12.1065Å Z=4	Pb(4b) O(16f) Mo(4a)	$Ca^{do}W^t[O_4]^*$ SCHEELITE	Zeit.Krist.,1965,121,158-164; SR,30A,328-329;Pov.,569;Str. Tab.,302;LF,150.
XENOTIME-(Y)	YPO_4	$Y^{do}P^t[O_4]^*$	Tet. I 4_1/amd	a=6.878Å Z=4 c=6.036Å	Y(4a) O(16h) P(4b)	$Zr^{do}Si^t[O_4]^*$ ZIRCON	Zeit.Krist.,1965,121,315-316; Pov.,536;SR,30A,386;LF,149.
ZIRCON	$ZrSiO_4$	$Zr^{do}Si^t[O_4]^*$	Tet. I 4_1/amd	a=6.607Å Z=4 c=5.982Å	Zr(4a) O(16h) Si(4b)	$Zr^{do}Si^t[O_4]^*$ ZIRCON	Am.Min.,1971,56,782-790; Pov.,381;LF,149.

Table 43

ABC_5

NAME	CHEMICAL FORMULA	STRUCTURAL FORMULA	SPACE GROUP	UNIT CELL DIMENSIONS	EQUIVALENT POSITIONS	STRUCTURE TYPE	REFERENCES
PALARSTANIDE	Pd_5SnAs		Trig.	a=7.50Å c=10.34Å Z=			Am.Min.,1989,74,1219-1220 (Abs.);Hölzel,11;Am.Min., 1982,67,858-859(Abs.).

Table 44

ABC_6

NAME	CHEMICAL FORMULA	STRUCTURAL FORMULA	SPACE GROUP	UNIT CELL DIMENSIONS	EQUIVALENT POSITIONS	STRUCTURE TYPE	REFERENCES
BURTITE	$CaSn(OH)_6$		Trig. $R\bar{3}$	a=11.49Å a_R=8.128Å c=14.08Å α≅90° Z=8 Z_R=4			Hölzel,84;Can.Min.,1981,19, 397-401.
JEANBANDYITE	(Fe,Mn)Sn $(OH)_6$		Tet. $P4_2/n$	a=7.648Å Z=4 c=7.648Å			Am.Min.,1983,68,471-472(Abs.); Hölzel,85.
MOPUNGITE	$NaSb(OH)_6$		Tet. $P4_2/n$	a=7.994Å Z=4 c=7.859Å			Min.Rec.,1985,16,73-74; Hölzel,85.
MUSHISTONITE	(Cu,Zn,Fe) $Sn(OH)_6$		Cub. $Pn\bar{3}m$	a=7.735Å Z=4			Am.Min.,1985,70,1331(Abs.); Hölzel,84.
NATANITE	$FeSn(OH)_6$		Cub. $Pn\bar{3}m$	a=7.69Å Z=4			Am.Min.,1982,67,1077(Abs.); Hölzel,84.
SCHOENFLIESITE	$MgSn(OH)_6$		Cub. $Pn\bar{3}$	a=7.759Å Z=4			Am.Min.,1972,57,1557-1558 (Abs.);Hölzel,84.
STOTTITE	$FeGe(OH)_6$	Fe^oGe^o $[\square_2(OH)_6]^c$ (Subs.deriv. Dzhalindite)	Tet. $P4_2/n$	a=7.594Å Z=4 c=7.488Å	Fe(4c) $(OH)_{I\text{-}III}$(8g) Ge(4d)		Am.Min.,1988,73,657-661(Abs.); Pov.,316;SR,23,399-400.
TETRAWICKMANITE	$MnSn(OH)_6$	Mn^oSn^o $[\square_2(OH)_6]^c$ (=Stottite)	Tet. $P4_2/n$	a=7.787Å Z=4 c=7.977Å			Am.Min.,1973,58,966-967(Abs.); Hölzel,85.
VISMIRNOVITE	$ZnSn(OH)_6$		Cub. $Pn\bar{3}m$	a=7.72Å Z=4			Am.Min.,1982,67,1077(Abs.); Hölzel,84.
WICKMANITE	$MnSn(OH)_6$	Mn^oSn^o $[\square_2(OH)_6]^c$ (Dist.deriv. Stottite)	Cub. $Pn\bar{3}m$	a=7.873Å Z=4			Am.Min.,1968,53,1063(Abs.); Pov.,316;Hölzel,84.

Table 45

AB_2C_2

NAME	CHEMICAL FORMULA	STRUCTURAL FORMULA	SPACE GROUP	UNIT CELL DIMENSIONS	EQUIVALENT POSITIONS	STRUCTURE TYPE	REFERENCES
BLIXITE	$Pb_2Cl(O,OH)_2$		Orth.	a=5.832Å Z=8 b=5.694Å c=25.47Å			Am.Min.,1960,45,908(Abs.); Pov.,654;Hölzel,56;Str.Tab.,169.
CHVILEVAITE	$Na(Cu,Fe,Zn)_2S_2$		Trig. P3m1	a=3.873Å Z=1 c=6.848Å			Am.Min.,1989,74,946(Abs.); Hölzel,15.
DERVILLITE	Ag_2AsS_2		Mon. P2/a	a=6.833Å β=99°33' b=12.932Å Z=2 c=9.638Å			Am.Min.,1983,68,1041(Abs.); Hölzel,35.
IMITERITE	Ag_2HgS_2	$Ag_2^{tt}Hg^{o}[S_2]^{h}$	Mon. $P2_1/c$	a=4.039Å β=107.12° b=8.0050Å Z=2 c=6.5812Å	Hg(2a) Ag(4e) S(4e)		Bull.Min.,1985,108,457-464; Am.Min.,1986,71,1277-1278 (Abs.);Hölzel,15.
KAWAZULITE	Bi_2Te_2Se	$Bi_2^{o}[Te_2Se]^{chh}$ or $Bi_{14}^{o}[Te_{13}Se_8]^{chh}$	Trig. $R\bar{3}m$...	a=4.24Å a_R=10.225° c=29.66Å α=24°5' Z=3 Z_R=1	Bi(6c) Te(6c) Se(3a)	$Bi_2^{o}[Te_2S]^{chh}$ or $Bi_{14}^{o}[Te_{13}S_8]^{chh}$ TETRADYMITE	Am.Min.,1972,57,1312(Abs.); Pov.,214-215;Wyckoff,2,29-31; LF,102.
KERMESITE	Sb_2OS_2		Tric. P1	a=11.66Å α=111°43' b=8.24Å β=92°40' c=11.19Å γ=110°44 Z=8	Sb_{I-IV}(8j) S_{I-IV}(8j) O_{I-II}(8f) (for C2/m)		Pov.,247;SR,32A,222;Hölzel, 29;Str.Tab.,129.
MELANOTHALLI-TE	Cu_2OCl_2		Orth. Fddd	a=9.595Å Z=8 b=9.693Å c=7.461Å			Am.Min.,1983,68,852(Abs.); Hölzel,53.
OREGONITE	$FeNi_2As_2$		Hex.	a=6.083Å Z=3 c=7.130Å			Am.Min.,1960,45,1130(Abs.); Hölzel,16;Pov.203-204.
PÄÄKKÖNENITE	Sb_2AsS_2		Mon.	a=5.372Å β=89.71° b=3.975Å Z=2 c=11.41Å			Am.Min.,1982,67,858(Abs.); Hölzel,suppl..
SINOITE	Si_2N_2O		Orth.	a=8.843Å Z=4 b=5.473Å c=4.835Å			Science,1964,146,256-257; RRW,563;Str.Tab.,97-98; Hölzel,5.
SKIPPENITE	Bi_2Se_2Te		Trig. $R\bar{3}m$	a=4.183Å Z=3 c=29.12Å			Can.Min.,1987,25,625-637; Hölzel,29.

Table 46

AB_2C_3

NAME	CHEMICAL FORMULA	STRUCTURAL FORMULA	SPACE GROUP	UNIT CELL DIMENSIONS	EQUIVALENT POSITIONS	STRUCTURE TYPE	REFERENCES
ARGENTOPYRITE	$AgFe_2S_3$	$Ag^{I}Fe_2^{I}[S_3]^{n}$	Orth. Pmmn	a=6.64Å Z=4 b=11.47Å c=6.45Å		$Cu^{I}Fe_2^{I}[S_3]^{n}$ CUBANITE	Am.Min.,1954,39,475-488; Pov.,226-227;Hölzel,22;Str. Tab.,120;LF,115.
ATACAMITE	$Cu_2Cl(OH)_3$	$Cu_2^{[6]}[Cl(OH)_3]^{c}$	Orth. Pnma	a=6.030Å Z=4 b=6.865Å c=9.120Å	$Cu_I(4a)$ $Cu_{II}(4c)$ Cl(4c) $O_I(4c)$ $O_{II}(8d)$ $H_I(4c)$ $H_{II}(8d)$		Acta Cryst.,1986,C42,1277-1280;Pov.,648;Hölzel,53.
BOTALLACKITE	$Cu_2Cl(OH)_3$	$Cu_2^{[6]}[Cl(OH)_3]^{c}$	Mon. $P2_1/m$	a=5.717Å β=93.07° b=6.126Å Z=2 c=5.636Å	$Cu_I(2e)$ $Cu_{II}(2d)$ Cl(2e) $O_I(2e)$ $O_{II}(4f)$ $H_I(2e)$ $H_{II}(4f)$		Min.Mag.,1985,49,87-89;Pov.,648;Sov.Phys. Cryst.,1958,3,445-451.
CUBANITE	$CuFe_2S_3$	$Cu^{I}Fe_2^{I}[S_3]^{n}$	Orth. Pcmn	a=6.467Å Z=4 b=11.117Å c=6.231Å	Cu(4c) Fe(8d) $S_I(4c)$ $S_{II}(8d)$	$Cu^{I}Fe_2^{I}[S_3]^{n}$ CUBANITE	Zeit.Krist.,1974,140,218-239; Pov.,226-227;Wyckoff,2,506-507;LF,115.
FISCHESSERITE	Ag_3AuSe_2	$Ag_3^{[4]}Au^{[2]}$ $[Se_2^{[8]}]^{c}$	Cub. I 432	a=9.967Å Z=8			Am.Min.,1972,57,1554(Abs,); Hölzel,14.
GREGORYITE	$(Na,K,Ca)_2CO_3$ α-form		Hex. $P6_3mc$	a=5.215Å Z=2 c=6.584Å			JCPDS,25-815;Hölzel,99.
ISOCUBANITE	$CuFe_2S_3$	$Cu^{I}Fe_2^{I}[S_3]^{c}$	Cub. $Fm\bar{3}m$	a=5.303Å Z=1			Min.Mag.,1988,52,509-514; Hölzel,23.
JALPAITE	Ag_3CuS_2		Tet. $I\,4_1/amd$	a=8.633Å c=11.743Å Z=8			Am.Min.,1968,53,1530-1542; Pov.,241;Hölzel,13;Str.Tab., 112.
KEMPITE	$Mn_2Cl(OH)_3$		Orth. Pnam	a=6.47Å Z=4 b=9.52Å c=7.12Å			RRW,321;Hölzel,53;Pov., 648.
MAGNOLITE	Hg_2TeO_3		Orth. Pbm2	a=5.958Å Z=2 b=10.576Å c=3.749Å	Hg(4d) Te(2c) $O_I(4d)$ $O_{II}(2c)$		Can.Min.,1989,27,133-136, 129-131;Hölzel,suppl..
MOHITE	Cu_2SnS_3	$Cu_2^{I}Sn^{I}[S_3]^{n}$	Tric. $P\bar{1}$	a=6.64Å α=90° b=11.51Å β=109°45' c=19.93Å γ=90° Z=12			Am.Min.,1983,68,281(Abs.); Hölzel,24;G.O.,Sér.Geol., 1986,,9,41-53.
NATRITE	Na_2CO_3 γ-form		Mon.	a=8.906Å β=101°21' b=5.238Å Z=4 c=6.045Å			Am.Min.,1983,68,281(Abs.); Hölzel,96.
PARATACAMITE	$Cu_2Cl(OH)_3$		Trig. $R\bar{3}$	a=13.654Å a_R=9.168Å c=14.041Å α=96.26° Z=24 Z_R=8	$Cu_I(3b)$ $Cu_{II}(9d)$ $Cu_{III-IV}(18f)$ $Cl_I(6c)$ $Cl_{II}(18f)$ $O_{I-IV}(18f)$		Acta Cryst.,1975,B31,183-187;SR,41A,195;Pov.,648; Str.Tab.,165.
PENFIELDITE	$Pb_2Cl_3(OH)$		Hex. P6/m…	a=11.28Å Z=36 c=48.65Å			RRW,469;Bull.Min.,1968,91, 407-408;Pov.,654.Hölzel,56.
PETZITE	Ag_3AuTe_2	$Ag_3^{[4]}Au^{[2]}[Te_2]^{c}$	Cub. $I\,4_132$	a=10.385Å Z=8	Ag(24f) Au(8a) Te(16e)		Min.Abs.,79-3411;SR,41A, 68;Pov.,212.

Table 47

AB_2C_3(cont.)

NAME	CHEMICAL FORMULA	STRUCTURAL FORMULA	SPACE GROUP	UNIT CELL DIMENSIONS	EQUIVALENT POSITIONS	STRUCTURE TYPE	REFERENCES
PICOTPAULITE	$TlFe_2S_3$	$Tl^tFe_2^t[S_3]^n$	Orth. C222...	a=5.40Å Z=4 b=10.72Å c=9.04Å			Am.Min.,1972,57,1909(Abs.); Pov.,226-227;Hölzel,19.
RADHAKRISHNAITE	$PbTe_3(Cl,S)_2$		Tet. P4/m...	a=5.71Å Z=1 c=3.77Å			Can.Min.,1985,23,501-506; Hölzel,47.
RASVUMITE	KFe_2S_3		Orth. Cmcm	a=9.049Å Z=4 b=11.019Å c=5.431Å	Fe(8e) K(4c) S_I(4c) S_{II}(8g)		Am.Min.,1980,65,477-482; SR,46A,93;Am.Min.,1979, 64,776-778;Pov.,755,226.
STERNBERGITE	$AgFe_2S_3$		Orth. Cmma	a=6.63Å Z=8 b=11.60Å c=12.68Å			Am.Min.,1969,54,1198-1201; Pov.,265;Str.Tab.,120.
SULPHOTSUMOITE	Bi_3Te_2S		Trig. $P\bar{3}m1$	a=4.316Å Z=2 c=23.43Å			Am.Min.,1983,68,1250(Abs.): Hölzel,29.
UYTENBOGAARDTITE	Ag_3AuS_2		Tet. $P4_122$...	a=9.76Å Z=8 c=9.78Å			Can.Min.,1978,16,651-657; Am.Min.,1982,67,1081(Abs.).
ZABUYELITE	Li_2CO_3		Mon. C2/c	a=8.356Å β=114.6° b=4.964Å Z=4 c=6.185Å			Am.Min.,1990,75,243-244 (Abs.);Hölzel,suppl..

Table 48

AB_2C_4

NAME	CHEMICAL FORMULA	STRUCTURAL FORMULA	SPACE GROUP	UNIT CELL DIMENSIONS	EQUIVALENT POSITIONS	STRUCTURE TYPE	REFERENCES
ARCANITE	K_2SO_4	$K_2^{[9/10]}\{g\}[S^tO_4]$	Orth. Pnam	a=7.476Å Z=4 b=10.071Å c=5.763Å	S(4c) K_{I-II}(4c) O_{I-II}(4c) O_{III}(8d)		Acta Cryst.,1972,B28,2845-2852;SB,2,86-87;Pov.,583; Str.Tab.,272.
BERTHIERITE	$FeSb_2S_4$		Orth. Pnam	a=11.423Å Z=4 a=14.174Å c=3.765Å	Fe(4c) Sb_{I-II}(4c) S_{I-IV}(4c)		Zeit.Krist.,1989,186,31-33; Am.Min.,1955,40,226-238; K/S,67-68;SR,19,418-419.
BRUNOGEIERITE	$Fe_2(Ge,Fe)O_4$		Cub. $Fd\bar{3}m$	a=8.409Å Z=8			Am.Min.,1973,58,348(Abs.); Hölzel,61.
CALCIBORITE	CaB_2O_4	$Ca^{[8]}\{1\infty\}[B^{tr}B^tO_4]$	Orth. Pccn	a=8.38 Z=8 b=13.82Å c=5.006Å	Ca(8e) B_{I-II}(8e) O_{I-IV}(8e)		SR,35A,283-284;Min.Abs., 81-2438;Pov.,480-481.
CARROLLITE	$Cu(Co,Ni)_2S_4$	$(Co,Ni)_2{}^oCu^t[S_4]^c$	Cub. $Fd\bar{3}m$	a=9.516Å Z=8		$Al_2Mg^t[O_4]^c$ SPINEL	Min.Mag.,1980,43,733-739; Pov.,223:RRW,110;LF,120.
CHROMITE	$FeCr_2O_4$	$Cr_2{}^oFe^t[O_4]^c$	Cub. $Fd\bar{3}m$	a=8.35Å Z=8	Fe(8a) O(32e) Cr(16d)	$Al_2{}^oMg^t[O_4]^c$ SPINEL	Wyckoff,3,75-79;Pov.,285; Str.Tab.,177;LF,120.
CHRYSOBERYL	$BeAl_2O_4$	$Al_2{}^oBe^t[O_4]^n$	Orth. Pnma	a=9.407Å Z=4 b=5.4781Å c=4.4285Å	Al_I(4a) Al_{II}(4c) Be(4c) O_{I-II}(4c) O_{III}(8d)	$Mg_2{}^oSi^t[O_4]^n$ FORSTERITE	Sov.Phys.Cryst.,1985,30, 277-279;Pov.,283;Wyckoff,3, 91;LF,122.
CHURSINITE	Hg_2AsO_4		Mon. $P2_1/c$	a=8.71Å β=128.27° b=5.08Å Z=6 c=15.66Å			Am.Min.,1985,70,871(Abs.); Acta Cryst.,1973,B29,1666-1669;Hölzel,145.
COCHROMITE	$(Co,Ni,Fe)(Cr,Al,Fe)_2O_4$	$(Cr,Al,Fe)_2{}^o(Co,Ni,Fe)^t[O_4]^c$	Cub. $Fd\bar{3}m$	a=8.292Å Z=8		$Al_2{}^oMg^t[O_4]^c$ SPINEL	Am.Min.,1980,65,811(Abs.); Hölzel,60.
COULSONITE	FeV_2O_4	$V_2{}^oFe^t[O_4]^c$	Cub. $Fd\bar{3}m$	a=8.297Å Z=8		$Al_2{}^oMg^t[O_4]^c$ SPINEL	Am.Min.,1962,47,1284-1291; Pov.,285;Str.Tab.,177;LF,120
CUPROIRIDSITE	$CuIr_2S_4$	$Ir_2{}^oCu^t[S_4]^c$	Cub. $Fd\bar{3}m$	a=9.92Å Z=8		$Al_2{}^oMg^t[O_4]^c$ SPINEL	Am.Min.,1986,71,1277(Abs.); Hölzel,28.
CUPRORHODSITE	$CuRh_2S_4$	$Rh_2{}^oCu^t[S_4]^c$	Cub. $Fd\bar{3}m$	a=9.88Å Z=8		$Al_2{}^oMg^t[O_4]^c$ SPINEL	Am.Min.,1986,71,1277(Abs.); Hölzel,28.
CUPROSPINEL	$(Cu,Mg)Fe_2O_4$	$Fe_2{}^o(Cu,Mg)^t[O_4]^c$	Cub. $Fd\bar{3}m$	a=8.369Å Z=8	(Cu,Mg)(8a) Fe(16d)O(32e)	$Al_2{}^oMg^t[O_4]^c$ SPINEL	Can.Min.,1973,11,1003-1007;Wyckoff,3,78-79.
DAUBRÉELITE	$FeCr_2S_4$	$Cr_2{}^oFe^t[S_4]^c$	Cub. $Fd\bar{3}m$	a=9.966Å Z=8		$Al_2{}^oMg^t[O_4]^c$ SPINEL	SR,11,289;Str.Tab.,127-128.
DONATHITE	$(Fe,Mg)(Cr,Fe)_2O_4$	$(Cr,Fe)_2{}^o(Fe,Mg)^t[O_4]^c$	Tet. $P4_2/nnm$	a=8.342Å c=8.305Å Z=8		Dist.deriv. $Al_2{}^oMg^t[O_4]^c$ SPINEL	Am.Min.,1969,54,1218-1219 (Abs.);Str.Tab.,178;Hölzel,,60.
FAYALITE	Fe_2SiO_4	$Fe_2{}^oSi^t[O_4]^n$	Orth. Pbnm	a=4.818Å Z=4 b=10.471Å c=6,086Å	Fe_I(4a) O_{I-II}(4c) Fe_{II}(4c) O_{III}(8d) Si(4c)	$Mg_2{}^oSi^t[O_4]^n$ FORSTERITE	Am.Min.,1975,60,1092-1097; Pov.,384-385;LF,122.
FILIPSTADITE	$(Mn,Mg)_2(Sb,Fe)O_4$	$(Mn,Mg)_2{}^o(Sb,Fe)^t[O_4]^c$	Cub.~	a=8.640Å Z=8		$Al_2{}^oMg^t[O_4]^c$ SPINEL~	Am.Min.,1988,73,413-419; Hölzel,64.

Table 49

AB_2C_4(cont.)

NAME	CHEMICAL FORMULA	STRUCTURAL FORMULA	SPACE GROUP	UNIT CELL DIMENSIONS	EQUIVALENT POSITIONS	STRUCTURE TYPE	REFERENCES
FLETCHERITE	$Cu(Ni,Co)_2S_4$	$(Ni,Co)_2{}^oCu^t[S_4]^c$	Cub. $Fd\bar{3}m$	a=9.520Å Z=8		$Al_2{}^oMg^t[O_4]^c$ SPINEL	Am.Min.,1977,62,1057(Abs.); Hölzel,27.
FLORENSOVITE	$Cu(Cr,Sb)_2S_4$		Cub. $Fd\bar{3}m$	a=10.005Å Z=8			Hölzel,suppl..
FORSTERITE	Mg_2SiO_4	$Mg_2{}^oSi^t[O_4]^h$	Orth. Pbnm	a=4.752Å Z=4 b=10.193Å (23°C) c=5.977Å	Mg_I(4a) Mg_{II}(4c) Si(4c) O_{I-II}(4c) O_{III}(8d)	$Mg_2{}^oSi^t[O_4]^h$ FORSTERITE	Am.Min.,1976,61,1280-1293; Pov.,384;Wyckoff,4,159-161; LF,122.
FRANKLINITE	$(Zn,Mn,Fe)(Fe,Mn)_2O_4$	$(Fe,Mn)_2{}^o(Zn,Mn,Fe)^t[O_4]^c$	Cub. $Fd\bar{3}m$	a=8.420Å Z=8		$Al_2{}^oMg^t[O_4]^c$ SPINEL	Am.Min.,1979,64,595-603: Pov.,285:LF.120.
GAHNITE	$ZnAl_2O_4$	$Al_2{}^oZn^t[O_4]^c$	Cub. $Fd\bar{3}m$	a=8.099Å Z=8	Zn(8a) Al(16d) O(32e)	$Al_2{}^oMg^t[O_4]^c$ SPINEL	Zeit.Krist.,1964,120,476-478;Pov.,284;LF,120.
GALAXITE	$MnAl_2O_4$	$Al_2{}^oMn^t[O_4]^c$	Cub. $Fd\bar{3}m$	a=8.287Å Z=8	Mn(8a) Al(16d) O(32e)	$Al_2{}^oMg^t[O_4]^c$ SPINEL	Am.Min.,1983,68,449455;Str. Tab.,177;Pov.,284;LF,120.
GALENOBISMUTITE	$PbBi_2S_4$	$Pb^{[7]}Bi_2{}^{[6/7]}[S_4]^{\#}$	Orth. Pnam	a=11.79Å Z=4 b=14.59Å c=4.10Å	Pb(4c) Bi_{I-II}(4c) Si_{I-IV}(4c)		Acta Cryst.,1962,15,691-698; Pov.,250;Wyckoff,3,134-136; LF,163.
GUETTARDITE	$Pb(Sb,As)_2S_4$		Mon. $P2_1/a$	a=20.05Å β=102°46' b=7.95Å Z=8 c=8.44Å			Can.Min.,1980,18,13-15; Pov.,253.
HEIDEITE	$(Fe,Cr)_{1+x}(Ti,Fe)_2S_4$	$(Ti,Fe)_2{}^o(Fe,Cr)_{1+x}{}^o[S_4]^h$	Mon. I 2/m	a=5.97Å β=90.2° b=3.42Å Z=2 c=11.4Å			Am.Min.,1974,59,465-470; Hölzel,28.
HERCYNITE	$FeAl_2O_4$	$Al_2{}^oFe^t[O_4]^c$	Cub. $Fd\bar{3}m$	a=8.136Å Z=8		$Al_2{}^oMg^t[O_4]^c$ SPINEL	RRW,269;Str,Tab.,177:LF,120.
HETAEROLITE	$ZnMn_2O_4$	$Mn_2{}^oZn^t[O_4]^c$	Tet. $I\,4_1/amd$	a=5.75Å Z=4 c=9.17Å		Dist.deriv. $Al_2{}^oMg^t[O_4]^c$ SPINEL	Pov.,287;Hölzel,61.
INDITE	$FeIn_2S_4$	$In_2{}^oFe^t[S_4]^c$	Cub. $Fd\bar{3}m$	a=10.618Å Z=8	$(Fe,In)^t$(8a) $(In,Fe)^o$(16d) S(32e)	$Al_2{}^oMg^t[O_4]^c$ SPINEL	SR,45A,79-80;Am.Min.,1964, 49,439440(Abs.);Pov.,223; LF,120.
IWAKIITE	$Mn(Fe,Mn)_2O_4$	$(Fe,Mn)_2{}^oMn^t[O_4]^c$	Tet. $I\,4_1/amd$	a=6.025Å Z=4 c=8.539Å		Dist.deriv. $Al_2{}^oMg^t[O_4]^c$ SPINEL	Zeit.Krist.,1988,185,605; Am.Min.,1980,65,406(Abs,).
JACOBSITE	$(Mn,Fe,Mg)(Fe,Mn)_2O_4$	$(Fe,Mn)_2{}^o(Mn,Fe,Mg)^t[O_4]^c$	Cub. $Fd\bar{3}m$	a=8.47Å Z=8		$Al_2{}^oMg^t[O_4]^c$ SPINEL	Pov.,285;Str.Tab.,177;RRW, 307;LF,120.
JOSÉITE-A	Bi_4TeS_2	$Bi_4{}^o[TeS_2]^{c/h}$	Trig. $R\bar{3}m$	a=4.24Å a_R=13.45Å c=39.69Å α=18°8' Z=3 Z_R=1			Am.Min.,1949,34,342-382; Pov.,214;RRW,307;LF,120.
JOSÉITE-B	Bi_4Te_2S	$Bi_4{}^o[Te_2S]^{c/h}$	Trig. $R\bar{3}m$	a=4.33Å a_R=13.81Å c=40.75Å α=18°2' Z=3 Z_R=1			Am.Min.,1949,34,342-382; Pov.,214;RRW,313.
KADYRELITE	$Hg_4(Br,Cl)_2O$		Cub. $I\,a\bar{3}d$	a=16.22Å Z=24			Am.Min.,1989,74,503(Abs.); Hölzel,55.

Table 50

AB_2C_4(cont.)

NAME	CHEMICAL FORMULA	STRUCTURAL FORMULA	SPACE GROUP	UNIT CELL DIMENSIONS	EQUIVALENT POSITIONS	STRUCTURE TYPE	REFERENCES
KALININITE	$ZnCr_2S_4$	$Cr_2{}^oZn^t[S_4]^c$	Cub. $Fd\bar{3}m$	a=9.997Å Z=8		$Al_2{}^oMg^o[O_4]^c$ SPINEL	Am.Min.,1987,72,223(Abs.); Hölzel,27.
LAIHUNITE	$Fe_{1.6}SiO_4$	$(Fe_{0.8}{}^{+2},Fe_{0.8}{}^{+3},\square_{0.4})^oSi^t[O_4]^c$	Mon. $P2_1/b$	a=4.805Å β=91.0° b=10.189 Å Z=12 c=17.403Å	$Fe_I(4a)$ $Fe_{II}(4c)$ Si(4c) $O_{I-IV}(4c)$	Dist.defect deriv. $Al_2{}^oMg^t[O_4]^c$ SPINEL	Am.Min.,1986,71,1455-1460; Am.Min.,1978,63,424-425 (Abs.).
LARNITE	Ca_2SiO_4 β-form		Mon. $P2_1/n$	a=5.48 Å β=94°53' b=6.76Å Z=4 c=9.28Å	$Ca_I(4e)$ $Ca_{II}(4e)$ Si(4e) $O_{I-IV}(4e)$		SR,16,335-336;Pov.,386;Str. Tab.,367;RRW,343.
LEITEITE	$ZnAs_2O_4$		Mon. $P2_1/c$	a=4.542 Å β=90.81° b=5.022 Å Z=4 c=17.597Å	$As_{I-II}(4e)$ Zn(4e) $O_{I-IV}(4e)$		Am.Min.,1987,72,629-632: Am.Min.,1977,62,1259-1262 (Abs.);Hölzel,91.
LIEBENBERGITE	$(Ni,Mg)_2SiO_4$	$(Ni,Mg)_2{}^oSi^t[O_4]^n$	Orth. Pbnm	a=4.7366Å Z=4 b=10.164Å c=5.9332Å	$(Ni,Mg)_I(4a)$ $(Ni,Mg)_{II}(4c)$ Si(4c) $O_{I-II}(4c)$ $O_{III}(8d)$	$Mg_2{}^oSi^t[O_4]^n$ FORSTERITE	Am.Min.,1981,66,770-776;Am. Min.,1973,58,733-735.
MAGNESIOCHROMI-TE	$MgCr_2O_4$	$Cr_2{}^oMg^t[O_4]^c$	Cub. $Fd\bar{3}m$	a=8.321Å Z=8	Mg(8a) Cr(16d) O(32e)	$Al_2{}^oMg^t[O_4]^c$ SPINEL	Str.Tab.,177;Wyckoff,3,78-79.
MAGNESIOFERRITE	$MgFe_2O_4$	$Fe_2{}^oMg^t[O_4]^c$	Cub. $Fd\bar{3}m$	a=8.383Å Z=8	Mg(8a) Fe(16d) O(32e)	$Al_2{}^oMg^t[O_4]^c$ SPINEL	Str.Tab.,177;Wyckoff,3,78-79; LF,120.
MANGANOCHROMI-TE	$(Mn,Fe)(Cr,V)_2O_4$	$(Cr,V)_2{}^o(Mn,Fe)^t[O_4]^c$	Cub. $Fd\bar{3}m$	a=8.47Å Z=8		$Al_2{}^oMg^t[O_4]^c$ SPINEL	Am.Min.,1978,63,1166-1174; Hölzel,60.
MAROKITE	$CaMn_2O_4$		Orth. Pmab ...	a=9.71Å Z=4 b=10.03Å c=3.162Å	Ca(4d) $O_I(4c)$ Mn(8e) $O_{II}(4d)$ $O_{III}(8e)$		SR,29,321;SR,31A,136;Bull. Min.,1966,89,318-324;Pov., 288.
MASCAGNITE	$(NH_4)_2SO_4$		Orth. Pmcn	a=5.98Å Z=4 b=10.62Å c=7.78Å			Pov.,583;Str.Tab.,273;Hölzel, 120.
PHENAKITE	Be_2SiO_4	$Be_4{}^tSi_2{}^t[O_8\square]^{(3R21)}$ by	Trig. $R\bar{3}$	a=12.45Å a_R=7.7Å0 c=8.23Å α=108°1' Z=18 Z_R=6	$Be_{I-II}(18f)$ Si(18f) $O_{I-IV}(18f)$		Pov.,383;Am.Min.,1987,72,769-777;Wyckoff,3,133-134;LF, 145.
PLATYNITE	$PbBi_2(Se,S)_4$		Trig. $R\bar{3}m$	a=8.49Å Z=8 c=20.80Å			Pov.,261;RRW,484;Hölzel,30.
QANDILITE	$(Mg,Fe)_2(Ti,Fe,Al)O_4$	$(Mg,Fe)_2{}^o(Ti,Fe,Al)^t[O_4]^c$	Cub. $Fd\bar{3}m$	a=8.4033Å Z=8		$Al_2{}^oMg^t[O_4]^c$ SPINEL	Min.Mag.,1985,49,739-744; Hölzel,suppl..
RINGWOODITE	$(Mg,Fe)_2SiO_4$		Cub.	a=8.122Å Z=8			Can.Min.,1977,15,79-101; Hölzel,186;RRW,519.
SARTORITE	$PbAs_2S_4$		Mon. $P2_1/n$	a=19.62Å β=90° b=7.89Å Z=4 c=4.19Å	Pb(4g) $As_{I-II}(4g)$ $S_{I-IV}(4g)$		Acta Cryst.,1961,14,1291 1292;Pov.,235;Str.Tab.,146; K/S,75.

Table 51

AB_2C_4(cont.)

NAME	CHEMICAL FORMULA	STRUCTURAL FORMULA	SPACE GROUP	UNIT CELL DIMENSIONS	EQUIVALENT POSITIONS	STRUCTURE TYPE	REFERENCES
SCHAFARZIKITE	$FeSb_2O_4$	$1\infty[Fe^oSb_2{}^{[3n]}O_4]$	Tet. $P4_2/mbc$	a=8.59Å Z=4 c=5.91Å	$Fe(4d)\ O_I(8g)$ $Sb(8h)\ O_{II}(8h)$		Am.Min.,1981,66,1073-1079; SR,41A,437;Pov.,306-307.
SIEGENITE	$CoNi_2S_4$	$Ni_2{}^oCo^t[S_4]^c$	Cub. $Fd\bar{3}m$	a=9.43Å Z=8		$Al_2{}^oMg^t[O_4]^c$ SPINEL	Str.Tab.,127;Pov.,223;Min.Abs., 80-4873;
SPINEL	$MgAl_2O_4$	$Al_2{}^oMg^t[O_4]^c$	Cub. $Fd\bar{3}m$	a=8.0800Å (26°C) Z=8	Mg(8a) Al(16d) O(32e)	$Al_2{}^oMg^t[O_4]^c$ SPINEL	Wyckoff,3,75-77;Pov.,284; Str.Tab.,176;LF,120.
TARAPACÁITE	K_2CrO_4		Orth. Pnam	a=7.663Å Z=4 b=10.388Å c=5.922Å	Cr(4c) $K_{I-II}(4c)$ $O_{I-II}(4c)\ O_{III}(8d)$		Acta Cryst.,1972,B28,2845- 2852;Pov.,575.
TEPHROITE	Mn_2SiO_4	$Mn_2{}^oSi^t[O_4]^n$	Orth. Pbnm	a=4.879Å Z=4 b=10.589Å c=6.234Å	$Mn_I(4a)$ $Mn_{II}(4c)\ O_{I-II}(4c)$ $Si(4c)\ O_{III}(8d)$	$Mg_2{}^oSi^t[O_4]^n$ FORSTERITE	Am.Min.,1980,65,1263-1269; SR,45A,386;Pov.,761,384.
THENARDITE	Na_2SO_4 α-form		Orth. Fddd	a=9.829Å Z=8 b=12.302Å c=5.868Å	Na(16g) S(8a) O(32h)		Can.Min.,1975,13,181-187; Pov.,582;Wyckoff,3,109-111; LF,310.
TREVORITE	$NiFe_2O_4$	$Fe_2{}^oNi^t[O_4]^c$	Cub. $Fd\bar{3}m$	a=8.43Å Z=8	Ni(8a) Fe(16d) O(32e)	$Al_2{}^oMg^t[O_4]^c$ SPINEL	Pov.,285;Wyckoff,3,78-79; RRW,624;LF,120.
TRIPPKEITE	$CuAs_2O_4$		Tet. $P4_2/mbc$	a=8.59Å Z=4 c=5.57Å	$Cu(4d)\ O_I(8g)$ $As(8h)\ O_{II}(8h)$		SR,41A,340;Str.Tab.,181;Pov., 306.
TWINNITE	$Pb(Sb,As)_2S_4$		Orth. Pnmn	a=19.6Å Z=8 b=7.99Å c=8.60Å			Can.Min.,1967,9,191-213; RRW,635:Pov.,235;Str.Tab., 149.
ULVÖSPINEL	Fe_2TiO_4	$Fe_2{}^oTi^t[O_4]^c$	Cub. $Fd\bar{3}m$	a=8.504Å Z=8	Ti(8a) Fe(16d) O(32e)	$Al_2{}^oMg^t[O_4]^c$ SPINEL	SR,46A,282;Pov.,285;Str.Tab., 177;LF,120.
VIOLARITE	$FeNi_2S_4$	$Ni_2{}^oFe^t[O_4]^c$	Cub. $Fd\bar{3}m$	a=9.53Å Z=8		$Al_2{}^oMg^t[O_4]^c$ SPINEL	Pov.,223;Str.Tab.,127;LF,120.
VOZHMINITE	$(Ni,Co)_4$ $(As,Sb)S_2$		Hex.	a=17.46Å Z=18 c=7.20Å			Am.Min.,1983,68,645(Abs.); Hölzel,16;
VUORELAINENITE	(Mn,Fe) $(V,Cr)_2O_4$	$(V,Cr)_2{}^o(Mn,Fe)^t$ $[O_4]^c$	Cub. $Fd\bar{3}m$	a=8.48Å Z=8		$Al_2{}^oMg^t[O_4]^c$ SPINEL	Can.Min.,1982,20,281-290; Hölzel, 61.
WADSLEYITE	$(Mg,Fe)_2SiO_4$ β-form		Orth. I mma	a=5.70Å Z=8 b=11.51Å c=8.24Å			Can.Min.,1983,21,29-35; Hölzel,186.Phys.Chem.Min., 1996,23,461-469.
WILLEMITE	Zn_2SiO_4	$Zn_4{}^tSi_2{}^t$ $[O_8\square]^{(3R21)}$ by	Trig. $R\bar{3}$	a=13.948Å a_R=8.64Å c=9.315Å α=107°46' Z=18 Z_R=6	$Zn_{I-II}(18f)$ Si(18f) $O_{I-IV}(18f)$		Acta Cryst.,1978,B34,3324- 3325;Pov.,383-384;Wyckoff,3, 133-134,LF,145;Str.Tab.,363.
ZINCOCHROMITE	$ZnCr_2O_4$	$Cr_2{}^oZn^t[O_4]^c$	Cub. $Fd\bar{3}m$	a=8.352Å Z=8	Zn(8a) Cr(16d) O(32e)	$Al_2{}^oMg^t[O_4]^c$ SPINEL	Am.Min.,1988,73,931(Abs.); Wyckoff,3,79;Pov.,765.

Table 52

AB_2C_5

NAME	CHEMICAL FORMULA	STRUCTURAL FORMULA	SPACE GROUP	UNIT CELL DIMENSIONS	EQUIVALENT POSITIONS	STRUCTURE TYPE	REFERENCES
ANDALUSITE	Al_2SiO_5	$Al^oAl^{[5]}Si^t$ $[O_5\square]^{R21\ my}$	Orth. Pnnm	a=7.7942Å Z=4 b=7.8985Å c=5.559Å	$Al^o(4e)$ $O_{I-III}(4g)$ $Al^{[5]}(4g)$ $O_{IV}(8h)$ Si(4g)	$Al^oAl^{[5]}Si^t$ $[O_5\square]^{R21\ my}$ ANDALUSITE	Wyckoff,4,188-189;Pov.,386; LF,144.
ARMALCOLI-TE	$(Mg,Fe)Ti_2O_5$	$(Mg,Fe)^{[6]}Ti^{[6]}$ $\{1\infty\}[Ti^oO_5]$	Orth. Bbmm	a=9.7762Å Z=4 b=10.0341Å c=3.7504Å	(Fe,Mg,Ti)(4c) (Ti,Fe,Mg)(8f) $O_I(4c)$ $O_{II-III}(8f)$	Dist.deriv. $Fe_2^{[6]}\{1\infty\}[Ti^oO_5]$ PSEUDOBROOKITE	Am.Min.,1977,62,913-920; Pov.,304-305;Am.Min.,1988, 73,1377-1383.
BERDESINS-KIITE	V_2TiO_5		Mon. C2/c...	a=10.11Å β=111.46° b=5.084Å Z=4 c=7.03Å			Am.Min.,1982,67,1074(Abs.); Hölzel,73,Am.Min.,1983,68, 1038(Abs.).
KYANITE	Al_2SiO_5	$Al_2{}^oSi^t[O_5]^c$	Tric. P1	a=7.1192Å α=89.977° b=7.8473Å β=101.121° c=5.5724Å γ=106.006° Z=4	$Al_{I-IV}(2i)$ $Si_{I-II}(2i)$ $O_{I-X}(2i)$	$Al_2{}^oSi^t[O_5]^c$ KYANITE	Wyckoff,4,180-183;Am.Min., 1979,64,573-586;Pov.,386; Str.Tab.,374;LF,127.
MULLITE	$Al_{4+x}Si_{2-2x}O_{10-x}$	$Al_2{}^o$ $\{1\infty\}[(Si_{2-2x},Al_{2+x})^t$ $O_{10-x}]$	Orth. Pbam	a=7.5786Å b=7.6817Å c=2.8864Å Z=1	Al(2a) $(Al,Si)_{I-II}(4h)$ $O_I(4h)$ $O_{II}(2d)$ $O_{III}(4h)$ $O_{IV}(4g)$ (part.occ.)	$Al_2{}^o$ $\{1\infty\}[(Si_{2-2x},Al_{2+x})^tO_{10-x}]$ MULLITE	Am.Min.,1986,71,1476-1482; Am.Min.,1972,62,247-255; Pov.,423-424;LF,216;SR, 31A,366-367;Wyckoff,4,187-
PSEUDO-BROOKITE	Fe_2TiO_5	$Fe_2^{[6]}\{1\infty\}[Ti^oO_5]$	Mon. C2	a=22.23Å β=116.2° b=3.73Å Z=4 c=9.80Å	Ti(4c) Fe(8f) $O_I(4c)$ $O_{II-III}(8f)$ (for Cmcm)	$Fe_2^{[6]}\{1\infty\}[Ti^oO_5]$ PSEUDOBROOKITE	Am.Min.,1988,73,1377-1383;SR,22,312-313;SR, 51A,208;Pov.,304-305
RAJITE	$CuTe_2O_5$		Mon. $P2_1/c$	a=6.865Å β=109.1° b=9.314Å Z=4 c=7.598Å			Min.Mag.,1979,43,91-92; Hölzel.94.
SANBORNITE	$BaSi_2O_5$	$Ba^{[9]}\{1\infty\}[Si_2{}^tO_5]$	Orth. Pmcn	a=7.688Å Z=4 b=4.629Å c=13.523Å	Ba(4c) Si(8d) $O_I(4c)$ $O_{II-III}(8d)$	$Ba^{[9]}\{1\infty\}[Si_2{}^tO_5]$ SANBORNITE	Zeit.Krist.,1980,153,33-41; Am.Min.,1958,44,517-536; Pov.,428;SR,22,490-491.
SILLIMANITE	Al_2SiO_5	$Al^o\{1\infty\}[Si^tAl^tO_5]$	Orth. Pbnm	a=7.4883Å Z=4 b=7.6808Å c=5.7774Å	$Al^o(4a)$ $O_{I-III}(4c)$ $Al^t(4c)$ $O_{IV}(8d)$ Si(4c)	$Al^o\{1\infty\}[Si^tAl^tO_5]$ SILLIMANITE	Am.Min.,1979,64,573-586; Wyckoff,4,186-187;Pov.,423-424;LF,216.
SMIRNITE	Bi_2TeO_5		Orth. Cm2a	a=16.447Å Z=8 b=5.513Å c=11.579Å			Am.Min.,1985,70,876-877 (Abs.);Hölzel,94.
STIBIVANITE	Sb_2VO_5		Mon. C2/c	a=17.989Å β=95.15° b=4.7924Å Z=4 c=5.500Å	Sb(8f) V(4e) $O_I(4e)$ $O_{II-III}(8f)$		Can.Min.,1980,18,333-337; Can.Min.,1980,18,329-332; Can.Min.,1989,27,625-632.
STIBIVANITE-2O	Sb_2VO_5		Orth. Pmcn	a=17.916Å Z=4 b=4.790Å c=5.509Å	Sb(8d) V(4c) $O_I(4c)$ $O_{II-III}(8d)$		Can.Min.,1989,27,625-632; Hölzel,suppl..

Table 53

AB_2C_6

NAME	CHEMICAL FORMULA	STRUCTURAL FORMULA	SPACE GROUP	UNIT CELL DIMENSIONS	EQUIVALENT POSITIONS	STRUCTURE TYPE	REFERENCES
AESCHYNITE-(Ce)	(Ce,Ca,Fe,Th) $(Ti,Nb)_2(O,OH)_6$	$(Ce,Ca,Fe,Th)^{[8]}$ $\{3\infty\}[(Ti,Nb)_2{}^{o}$ $(O,OH)_6]$	Orth. Pnma	a=11.08Å Z=4 b=7.56Å c=5.37Å	(Ce,Ca,Fe,Th)(4c) (Ti,Nb)(8d) O_{I-II}(8d) O_{III-IV}(4c)		SR,27,535-536;Pov.,296; Hölzel,76.
AESCHYNITE-(Y)	(Y,Ca,Fe,Th) $(Ti,Nb)_2(O,OH)_6$	$(Y,Ca,Fe,Th)^{[8]}$ $\{3\infty\}[(Ti,Nb)_2{}^{o}$ $(O,OH)_6]$	Orth. Pbnm	a=5.19Å Z=4 b=10.98Å c=7.40Å			Str.Tab.,208;Hölzel,76; RRW,4.
BARARITE	$(NH_4)_2SiF_6$	$Si^{o}[(NH_4)_2F_6]^{c}$	Trig. $C\bar{3}m$	a=5.76Å c=4.77Å Z=1	Si(1a) F(6i) (NH_4)(2d)	Dist.deriv. $Si^{o}[K_2F_6]^{c}$ HIERATITE	SB,3,123;Wyckoff,3,339-342;LF,106.
BRANNERITE	(U,Ca,Y,Ce) $(Ti,Fe)_2O_6$		Mon. C2/m	a=9.8123Å β=118.95° b=3.7697Å Z=2 c=6.9253Å	U(2a) Ti(4i) O_{I-III}(4i)		Can.Min.,1982,20,271-279; SR,31A,139;Pov.,311-312; Hölzel,78.
BYSTRÖMITE	$MgSb_2O_6$	$Mg^{o}Sb_2{}^{o}[O_6]^{h}$	Tet. $P4_2/mnm$	a=4.68Å c=9.21Å Z=2		$(Fe,Mn)^{o}(Ta,Nb)_2{}^{o}$ $[O_6]^{h}$ TAPIOLITE	Am.Min.,1952,37,53-57; Pov.,301;Str.Tab.,202; Wyckoff,3,361-362;LF,91.
CHANGBAIITE	$PbNb_2O_6$		Trig. R3m	a=10.499Å Z=9 c=11.553Å			Am.Min.,1979,64,242(Abs.); Hölzel,78.
CRYPTOHALITE	$(NH_4)_2SiF_6$	$Si^{o}[(NH_4)_2F_6]^{c}$	Cub. $Fm\bar{3}m$	a=8.395Å Z=4	Si(4a) F(24e) (NH_4)(8c)	$Si^{o}[K_2F_6]^{c}$ HIERATITE	Wyckoff,3,339-342;Pov.,665; Str.Tab.,160,43.
EUXENITE-(Y)	(Y,Ca,Ce,U,Th) $(Nb,Ta,Ti)_2O_6$	$(Y,Ca,Ce,U,Th)^{o}$ $(Nb,Ta,Ti)_2{}^{o}[O_6]^{h}$	Orth. Pbcn	a=14.643Å Z=4 b=5.553Å c=5.195Å	(Y,Ca,Ce,U,Th)(4c) (Nb,Ta,Ti)(8d) O_{I-III}(8d)	$(Fe,Mn)^{o}Ta_2{}^{o}[O_6]^{h}$ TANTALITE	Zeit.Krist.,1980,152,69-82; SR,46A,251;Pov.,310-311; Can.Min.,1977,15,92-96.
FERROCOLUMBI-TE	$FeNb_2O_6$	$Fe^{o}Nb_2{}^{o}[O_6]^{h}$	Orth. Pbcn	a=5.73Å Z=4 b=14.24Å c=5.08Å		$(Fe,Mn)^{o}Ta_2{}^{o}[O_6]^{h}$ TANTALITE	Hölzel,76;Str.Tab.,526.
FERROTAPIOLI-TE	$Fe(Ta,Nb)_2O_6$	$Fe^{o}(Ta,Nb)_2{}^{o}[O_6]^{h}$	Tet. $P4_2/mnm$	a=4.762Å Z=2 c=9.272Å		$(Fe,Mn)^{o}(Ta,Nb)_2{}^{o}$ $[O_6]^{h}$ TAPIOLITE	Am.Min.,1985,70,217(Abs.); Hölzel,73.
FERSMITE	(Ca,Ce,Na) $(Nb,Ta,Ti)_2$ $(O,OH,F)_6$	$(Ca,Ce,Na)^{o}$ $(Nb,Ta,Ti)_2{}^{o}$ $[O,OH,F]_6{}^{h}$	Orth. Pcan	a=5.718Å Z=4 b=14.91Å c=5.221Å	(Ca,Ce,Na)(4c) (Nb,Ta,Ti)(8d) O_{I-III}(8d)	$(Fe,Mn)^{o}Ta_2{}^{o}[O_6]^{h}$ TANTALITE	SR,24,367-368;Pov.,310-311;Am.Min.,1947,32,373 (Abs.).
FOORDITE	$Sn(Nb,Ta)_2O_6$	$Sn^{[8ap]}(Nb,Ta)_2{}^{o}$ $[O_6]^{h}$	Mon. C2/c	a=17.093Å β=90.85° b=4.877Å Z=4 c=5.558Å	Sn(4e) (Nb,Ta)(8f) O_{I-III}(8f)		Can.Min.,1988,26,899-903; Hölzel,suppl..
HIERATITE	K_2SiF_6	$Si^{o}[K_2F_6]^{c}$	Cub. $Fm\bar{3}m$	a=8.133Å Z=4	Si(4a) K(8c) F(24e)	$Si^{o}[K_2F_6]^{c}$ HIERATITE	Wyckoff,3,339-342;Pov.,665; Str.Tab.,160;LF,106.
KOECHLINITE	Bi_2MoO_6		Orth. $Pna2_1$	a=5.4822Å Z=4 b=16.1926Å c=5.5091Å	Bi_{I-II}(4a) Mo(4a) O_{I-VI}(4a)		Acta Cryst.,1984,C40,2001-2005;Pov.,312-313;SR,20, 449-450;SR,37A,240.
LAUTARITE	$Ca(IO_3)_2$		Mon. $P2_1/n$	a=7.143Å β=106.35° b=11.297Å Z=4 c=7.2804Å	Ca(4e) I_{I-II}(4e) O_{I-VI}(4e)		Acta Cryst.,1978,B34,84-88;Pov.,629;SR,44A,293-294.
LUCASITE-(Ce)	$(Ce,La)Ti_2$ $(O,OH)_6$		Mon. I 2/a	a=5.178Å β=93.52° b=8.756Å Z=4 c=9.768Å	(Ce,La)(4e) Ti(8f) $(O,OH)_{I-III}$(8f)		Am.Min.,1987,72,1006-1010; Hölzel,78.

Table 54

AB_2C_6(cont.)

NAME	CHEMICAL FORMULA	STRUCTURAL FORMULA	SPACE GROUP	UNIT CELL DIMENSIONS	EQUIVALENT POSITIONS	STRUCTURE TYPE	REFERENCES
LUDLOCKITE	$(Fe,Pb)As_2O_6$		Tric. P1...	a=10.41Å α=113.9° b=11.95Å β=99.7° c=9.86Å γ=82.7° Z=9			Am.Min.,1972,57,1003-1004(Abs.);Pov.,289;Hölzel,91.
MAGNOCOLUMBITE	(Mg,Fe,Mn) $(Nb,Ta)_2O_6$	$(Mg,Fe,Mn)^o$ $(Nb,Ta)_2{}^o[O_6]^h$	Orth. Pcan ?	a=5.02Å Z=4 b=14.17Å c=5.65Å		$(Fe,Mn)^oTa_2{}^o[O_6]^h$ TANTALITE	Am.Min.,1963,48,1182 (Abs.);Pov.,303;Str.Tab.,206.
MALLADRITE	Na_2SiF_6	$Si^o[Na_2F_6]^c$	Trig. P'321	a=8.859Å c=5.038Å Z=3	Si_I(1a) Na_I(3e) Si_{II}(2d) Na_{II}(3f) F_{I-III}(6g)	Dist.deriv. $Si^o[K_2F_6]^c$ HIERATITE	Acta Cryst.,1964,17,1408-1412;Pov.,668;Wyckoff,3,353-354.
MANGANOCOLUM-BITE	(Mn,Fe) $(Nb,Ta)_2O_6$	$(Mn,Fe)^o$ $(Nb,Ta)_2{}^o[O_6]^c$	Orth. Pbcn	a=14.42Å Z=4 b=5.78Å c=5.09Å		$(Fe,Mn)^oTa_2{}^o[O_6]^h$ TANTALITE	Str.Tab.,205-206;Pov.,303;Hölzel,76.
MANGANOTANTA-LITE	(Mn,Fe) $(Ta,Nb)_2O_6$	$(Mn,Fe)^o$ $(Ta,Nb)_2{}^o[O_6]^h$	Orth. Pbcn	a=14.413Å Z=4 b=5.760Å c=5.084Å	(Mn,Fe)(4c) Ta(8d) O_{I-III}(8d)	$(Fe,Mn)^oTa_2{}^o[O_6]^h$ TANTALITE	Can.Min.,1976,14,540-549;SR,42A,264;Pov.,303.
MANGANOTAPIO-LITE	(Mn,Fe) $(Ta,Nb)_2O_6$	$(Mn,Fe)^o$ $(Ta,Nb)_2{}^o[O_6]^h$	Tet. $P4_2/mnm$	a=4.762Å Z=2 c=9.272Å		$(Fe,Mn)^o(Ta,Nb)_2{}^o[O_6]^h$ TAPIOLITE	Am.Min.,1985,70,217 (Abs.);Pov.,301-302.
NIOBO-AESCHYNI-TE-(Ce)	(Ce,Ca) $(Nb,Ti)_2(O,OH)_6$	$(Ce,Ca)^{[8]}$ $(Nb,Ti)_2{}^o$ $[(O,OH)_6]$	Orth. Pbnm	a=5.396Å Z=4 b=11.085Å c=7.585Å		Dist.deriv. $(Fe,Mn)^o(Ta,Nb)_2{}^o[O_6]^h$ TAPIOLITE	Am.Min.,1975,60,309-315;Pov.,296;Str.Tab.,208.
NITROBARITE	$Ba(NO_3)_2$		Cub. $P2_13$	a=8.11Å Z=4	Ba(4a)O_{I-II}(12b) N_{I-II}(4a)		Zeit.Krist.,1967,124,310-334;Hölzel,96;Pov.,632.
ORDOÑEZITE	$ZnSb_2O_6$	$Zn^oSb_2{}^o[O_6]^h$	Tet. $P4_2/mnm$	a=4.67Å Z=2 c=9.24Å		$(Fe,Mn)^o(Ta,Nb)_2{}^o[O_6]^h$ TAPIOLITE	Pov.,301;Str.Tab.,202;Am.Min.,1955,40,64-69.
POLYCRASE-(Y)	$Y(Ti,Nb)_2$ $(O,OH)_6$		Orth. Pbcn	a≅14.57Å Z=4 b≅5.52Å c≅5.17Å			Str.Tab.,206;Pov.,310.
RUSSELLITE	Bi_2WO_6		Orth. $Pca2_1$	a=5.43726Å Z=4 b=16.43018Å c=5.4584Å			Min.Mag.,1992,56,399-409. Min.Mag.,1970,37,705-707;SB,7,108-109;Hölzel,79.
RYNERSONITE	$Ca(Ta,Nb)_2O_6$		Orth. Pmnb	a=7.505Å Z=4 b=11.063Å c=5.370Å			Am.Min.,1978,63,709-714;Hölzel,77.
TANTAL-AESCHYNI-TE-(Y)	(Y,Ce) $(Ta,Ti,Nb)_2O_6$		Orth.	a=5.34Å Z=1 b=10.97Å c=7.38Å (heated)			Min.Mag.,1974,39,571-576;Hölzel,76.
TANTALITE	$(Fe,Mn)Ta_2O_6$	$(Fe,Mn)^oTa_2{}^o$ $[O_6]^h$	Orth. Pbcn	a=14.413Å Z=4 b=5.760Å c=5.084Å	(Fe,Mn)(4c) Ta(8d)O_{I-III}(8d)	$(Fe,Mn)^oTa_2{}^o[O_6]^h$ TANTALITE	Can.Min.,1976,14,540-549;RRW,602;Wyckoff,3,362-364;Str.Tab.,205.

Table 55

AB_2C_6(cont.)

NAME	CHEMICAL FORMULA	STRUCTURAL FORMULA	SPACE GROUP	UNIT CELL DIMENSIONS	EQUIVALENT POSITIONS	STRUCTURE TYPE	REFERENCES
TANTEUXENI-TE-(Y)	(Y,Ce,Ca) $(Ta,Nb,Ti)_2$ $(O,OH)_6$	$(Y,Ce,Ca)^o$ $(Ta,Nb,Ti)_2{}^o$ $[(O,OH)_6]^h$	Orth. Pbcn	a=14.57Å Z=4 b=5.56Å c=5.18Å		$(Fe.Mn)^oTa_2{}^o[O_6]^h$ TANTALITE	Str.Tab.,206;Hölzel,76; Pov.,760,310.
TAPIOLITE	(Fe,Mn) $(Ta,Nb)_2O_6$	$(Fe,Mn)^o$ $(Ta,Nb)_2{}^o[O_6]^h$	Tet. $P4_2/mnm$	a=4.67Å Z=2 c=9.14Å	(Fe,Mn)(2a) (Ta,Nb)(4e) O_I(4f) O_{II}(8j)	$(Fe,Mn)^o(Ta,Nb)_2{}^o[O_6]^h$ TAPIOLITE	Wyckoff,3,361-362;Pov., 301;SR,33A,343-344:Am. Min.,1985,70,217(Abs.).
THOREAULITE	$SnTa_2O_6$	$Sn^{[8ap]}Ta_2{}^o[O_6]^h$	Mon. C2/c	a=17.140Å β=91.0° b=4.865Å Z=4 c=5.548Å	Sn(4e) Ta(8f) O_I(4e) O_{II-IV}(8f) (for $SnTa_2O_7$)		Am.Min.,1970,55,367-377; Am.Min.,1974,59,1026-1039; Pov.,304;SR,35A,444-445; Str.Tab.,211.
THORUTITE	(Th,U,Ca) $Ti_2(O,OH)_6$		Mon. C2/m	a=9.822Å β=108.84° b=3.824Å Z=2 c=7.036Å (synthetic)	Th(2a) Ti(4i) O_{I-III}(4i)		Acta Cryst.,1966,21,974-978;Pov.,311-312;Am.Min., 1958,43,1007(Abs.).
TRIPUHYITE	$FeSb_2O_6$	$Fe^oSb_2{}^o[O_6]^h$	Tet. $P4_2/mnm$	a=4.63Å Z=2 c=9.14Å		$(Fe,Mn)^o(Ta,Nb)_2{}^o[O_6]^h$ TAPIOLITE	Min.Mag.,1953,30,100-112; Pov.,301;Str.Tab.,202;LF,91.
VIGEZZITE	(Ca,Ce) $(Nb,Ta,Ti)_2O_6$		Orth. Pmnb...	a=7.559Å Z=4 b=11.028Å c=5.360Å			Min.Mag.,1979,43,459-462; Am.Min.,1980,65,811-812 (Abs.).
YTTROTANTALI-TE-(Y)	(Y,U,Ca) $(Ta,Fe)_2(O,OH)_6$		Orth.				Str.Tab.,206;Hölzel,77;Pov., 765,274.

Table 56

AB_2C_7

NAME	CHEMICAL FORMULA	STRUCTURAL FORMULA	SPACE GROUP	UNIT CELL DIMENSIONS	EQUIVALENT POSITIONS	STRUCTURE TYPE	REFERENCES
BARIOMICROLITE	$Ba(Ta,Nb)_2$ $(O,OH)_7$	$(Ba\square)^{cb}(Ta,Nb)_2{}^{[6]}$ $[(O,OH)_7\square]^{Qs}$	Cub. $Fd\bar{3}m$	a=10.570Å Z=8			Am.Min.,1977,62,403-410; Hölzel,67;LF,140.
BISMUTOSTIBICO-NITE	$Bi(Sb,Fe)_2O_7$	$(Bi\square)^{cb}(Sb,Fe)_2{}^{[6]}$ $[O_7\square]^{Qs}$	Cub. $Fd\bar{3}m$	a=10.38Å Z=8			Am.Min.,1984,69,1190-1196; Hölzel,66.
PLUMBOMICROLITE	(Pb,Na,Ca) $(Ta,Nb)_2(O,OH)_7$	$((Pb,Na,Ca)\square)^{cb}$ $(Ta,Nb)_2{}^{[6]}[(O,OH)_7\square]^{Qs}$	Cub. $Fd\bar{3}m$	a=10.58Å Z=8			Am.min.,1962,47,1220-1221 (Abs.);Pov.,276;Str.Tab.,192; Hölzel,67.
YTTROPYROCHLO-RE-(Y)	(Y,Ce,Nd,Th) $(Nb,Ta)_2(O,OH)_7$		(amorf, metamict.)	(a=10.3Å) (Z=8)(heated)			Am.Min.,1958,43,797;Hölzel, 67.

Table 57

AB_2C_8

NAME	CHEMICAL FORMULA	STRUCTURAL FORMULA	SPACE GROUP	UNIT CELL DIMENSIONS	EQUIVALENT POSITIONS	STRUCTURE TYPE	REFERENCES
FROLOVITE	$Ca(B(OH)_4)_2$		Tric. P¯1	a=7.774Å α=113.15° b=5.680Å β=101.67° c=8.136Å γ=107.87Å Z=2			Min.Abs.,78-262;Pov.,471.
LIANDRATITE	$U(Nb,Ta)_2O_8$		Trig. P$\bar{3}$1m	a=6.36Å Z=1 c=4.01Å			Am.Min.,1978,63,941-946; Hölzel,78.
PROSOPITE	$CaAl_2(F,OH)_8$		Mon. C2/c	a=6.70Å β=95.0° b=11.13Å Z=4 c=7.33Å	Ca(4e) (F,O)(8f) Al_I(4d) F_{II-III}(8f) Al_{II}(4e) O_{IV}(8f)		SR,41A,413;Pov.,653.
SEDOVITE	$U(MoO_4)_2$		Orth.	a=3.36Å Z=1 b=11.08Å c=6.42Å			Am.Min.,1966,51,530(Abs.); Pov.,569;Hölzel,141;Str.Tab., 302.

Table 58

$AB_mC_n(m,n\geq3)$

NAME	CHEMICAL FORMULA	STRUCTURAL FORMULA	SPACE GROUP	UNIT CELL DIMENSIONS	EQUIVALENT POSITIONS	STRUCTURE TYPE	REFERENCES
ANDESINE	$(Na,Ca)(Si,Al)_4O_8$	$(Na,Ca)^{[8]}\{3\infty\}[Si_2(Si,Al)^t Al^tO_8]$	Tric. P1	a=8.151Å α=93.62° b=12.829Å β=116.21° c=14.206Å γ=89.70° Z=8		$(Na,Ca)^{[8]}\{3\infty\}[Si_2(Si,Al)^tAl^tO_8]$ PLAGIOCLASE	Zeit.Krist.,1987,178,207 (Abs.);Pov.,344-345;Str. Tab.,478;LF,269.
ARGENTO-PENTLANDITE	$Ag(Fe,Ni)_8S_8$	$Ag^t(Fe,Ni)_8{}^t[S_8]^c$	Cub. $Fm\bar{3}m$	a=10.521Å Z=4	Ag(4b) S_I(24e) (Fe,Ni)(32f)S_{II}(8c)	$(Ni,Fe)^o(Ni,Fe)_8{}^t[S_8]^c$ PENTLANDITE	Can.Min.,1973,12,169-177; Hölzel,16.
ARGYRODITE	Ag_8GeS_6	$Ag_3{}^tAg_4{}^{tr}Ag^{[2]}Ge^t[S_6\square_2]^c$	Orth. $Pna2_1$	a=15.149Å Z=4 b=7.476Å c=10.589Å	Ag_{I-VIII}(4a) Ge(4a) S_{I-VI}(4a)	$Ag_3{}^tAg_4{}^{tr}Ag^{[2]}Ge^t[S_6\square_2]^c$ ARGYRODITE	Mon.Chem.,1977,108,901-913;SR,43A,66;G.O,Sér.Geol., 1986,9,41-53.
ARSENOSULVA-NITE	$Cu_3(As,V)S_4$	$Cu_3{}^t(As,V)^t[S_4]^c$ (Subs.deriv. Sphalerite)	Cub. $P\bar{4}3m$	a=5.3912Å Z=1	Cu(3d) S(4e) (As,V)(1a)		Am.Min.,1966,51,890-894; SR,31A,41;Str.Tab.,117;Pov., 230.
BANNERMANITE	$(Na,K)_xV_6O_{15}$		Mon. C2/m	a=15.413Å β=109.29° b=3.615Å Z=2 c=10.066Å	(Na,K)(4i) V_{I-III}(4i) O_I(2a) (part.occ.) $O_{II-VIII}$(4i)		Am.Min.,1983,68,634-641; Hölzel,87;Am.Min.,1990,75, 508-521.
BERNARDITE	$TlAs_5S_8$	$Tl^{[8]}\{3\infty\}[As_5S_8]$	Mon. $P2_1/c$	a=15.647Å β=91.27° b=8.038Å Z=4 c=10.750Å	Ti(4e) S_{I-VIII}(4e) As_{I-V}(4e)		Min.Mag.,1989,53,531-538; Hölzel,40.
BETEKHTINITE	$Cu_{10}(Pb,Fe)S_6$		Orth. Immm	a=3.86Å Z=4 b=14.67Å c=22.8Å	Pb(4g) ... Cu_{I-II}(4j) (Z=2 for $Pb_2(Cu,Fe)_{21}S_{15}$)		Acta Cryst.,1959,12,646-651; K/S,20;SR,23,140-141;Pov., 258;Hölzel,12.
BILLINGSLEYI-TE	$Ag_7(Sb,As)S_6$		Cub. $P2_13$	a=10.481Å Z=4			Can.Min.,1990,28,751-755. Am.Min.,1968,53,1791-1798; Pov.,231;Hölzel,38.
BORNITE	Cu_5FeS_4	$(Cu_5Fe\square_2)^t[S_4]^c$ (Defect deriv. Digenite)	Tet. $P\bar{4}2_1c$	a=10.94Å c=21.88Å Z=16			Am.Min.,1961,46,1270-1282; Pov.,231-232;Acta Cryst., 1975,B31,2268-2273;SR,44A, 56;SR,13,280-281;LF,109.
BOROVSKITE	Pd_3SbTe_4		Cub. $Fm\bar{3}m$?	a=5.794Å Z=1			Am.Min.,1974,59,873(Abs.); K/S,158-159;Hölzel,12.
BRAUNITE	Mn_7SiO_{12}	$Mn^{cb}Mn_6^{[6]}Si^t[O_{12}\square_4]^{cs}$	Tet. $I4_1/acd$	a=9.408Å c=18.668Å Z=8	Mn^{cb}(8b) Si(8b) $Mn^{[6]}_I$(16c) $Mn^{[6]}_{II}$(16e) ...	$Mn^{cb}Mn_6^{[6]}Si^t[O_{12}\square_4]^{cs}$ BRAUNITE	Am.Min.,1976,61,1226-1240; Pov.,287-288;Str.Tab.,377; LF,141.
BYTOWNITE	$(Ca,Na)(Si,Al)_4O_8$	$(Ca,Na)^{[8]}\{3\infty\}[Si_2{}^tAl_2{}^tO_8]$	Tric. $P\bar{1}$	a=8.178Å α=93°30' b=12.870Å β=115°54' c=14.187Å γ=90°39' Z=8	$(Ca,Na)_{I-II}$(2i) $(Si,Al)_{I-VIII}$(2i) O_{I-XVI}(2i)(1/2un.c.)	$Ca^{[8]}\{3\infty\}[Si_2{}^tAl_2{}^tO_8]$ ANORTHITE	Acta Cryst.,1966,21,782-801; SR,31A,222-223;Str.Tab.,475; Wyckoff,4,447-449;LF,269.
CALCIOTANTITE	$CaTa_4O_{11}$		Hex. $P6_3$...	a=6.23Å Z=2 c=12.22Å			Am.Min.,1983,68,471(Abs.); Hölzel,77.
CAMERONITE	$AgCu_7Te_{10}$		Tet. $P4_2mmc$...	a=12.695Å Z=16 c=42.186Å			Can.Min.,1986,24,379-384; Hölzel,14.
CANFIELDITE	Ag_8SnS_6	$Ag_3{}^tAg_4{}^{tr}Ag^{[2]}Sn^t[S_6\square_2]^c$	Orth. $Pna2_1$	a=15.133Å Z=4 b=7.463Å c=10.583Å		$Ag_3{}^tAg_4{}^{tr}Ag^{[2]}Ge^t[S_6\square_2]^c$ ARGYRODITE	N.Jb.Min.Mh.,1978,269- ; Hölzel,14;Pov.,731,231;G.O. Ser.Geol.,1986,9,41-53.

Table 59

$AB_mC_n(m,n)\geq 3$(cont.)

NAME	CHEMICAL FORMULA	STRUCTURAL FORMULA	SPACE GROUP	UNIT CELL DIMENSIONS	EQUIVALENT POSITIONS	STRUCTURE TYPE	REFERENCES
CARLFRIESITE	$CaTe_3O_8$		Mon. C2/c	a=12.576Å β=115.56° b=5.662Å Z=4 c=9.994Å	Te_I(8f) Ca(4e) Te_{II}(4a) $O_{I\text{-}IV}$(8f)		Am.Min.,1978,63,847-852; SR,44A,290;Min.Mag.,1975, 40,127-130;Hölzel.
CHAMÉANITE	$(Cu,Fe)_4As$ $(Se,S)_4$		Cub.	a=11.039Å Z=8			Am.Min.,1982,67,1074-1075; (Abs.);Hölzel,52.
CHLORMANGA-NOKALITE	K_4MnCl_6		Trig. $R\bar{3}c$	a=11.93Å a_R=8.468Å c=14.81Å α=89°32' Z=6 Z_R=2	Mn(2a) K_I(2b) K_{II}(6d) Cl(12f)(rhom.d.)		SR,11,413-415;Pov.,641; Hölzel,52.
CLIFFORDITE	UTe_3O_9		Cub. $Pa\bar{3}$	a=11.370Å Z=8	U_I(4a) $O_{I\text{-}II}$(24d) U_{II}(4b) $O_{III\text{-}V}$(8c) Te(24d)		Acta Cryst.,1971,B27,608- 616;SR,37A,317-318;Pov., 564;Am.Min.,1969,54,697-701.
CORONADITE	$PbMn_8O_{16}$	$Mn_8{}^o[PbO_{16}]^{chh}$	Mon. I 2/m	a=9.938Å β=90.39° b=2.8678Å Z=1 c=9.834Å	$Mn_{I\text{-}II}$(4i) $O_{I\text{-}IV}$(4i) Pb(4g) (occ.1.07)	$Mn_8{}^o[(Ba,K)O_{16}]^{chh}$ HOLLANDITE	Am.Min.,1989,74,913-917; Pov.,305;Hölzel,73;LF,107.
CRICHTONITE	(Sr,La,Ce) $(Ti,Fe,Mn)_{21}O_{38}$	$(Ti,Mn)_{19}{}^oFe_2{}^t$ $[(Sr,La,Ce)O_{38}]^{hhc}$	Trig. $R\bar{3}$	a=10.374Å a_R=9.148Å c=20.746Å α=69.076° Z=3 Z_R=1	(Sr,La,Ce)(1a) Fe(2c) (Ti,Fe,Mn)(1b)...		Am.Min.,1976,61,1203-1212; Pov.,280;SR,42A,254-255.
CROOKESITE	Cu_7TlSe_4		Tet. $I\bar{4}$	a=10.4456Å Z=2 c=3.9310Å			Zeit.Krist.,1987,181,241-249; Wyckoff,3,295-296;K/S,170-174
CRYOLITE	Na_3AlF_6 α-form	$Na^oAl^o[Na_2F_6]^c$	Mon. $P2_1/n$	a=5.46Å β=90°11' b=5.61Å Z=2 c=7.80Å	Al(2a) $F_{I\text{-}III}$(4c) Na_I(2b) Na_{II}(4c)	Dist.deriv. $Al^oNa^o[K_2F_6]^c$ ELPASOLITE	Wyckoff,3,382-383;Pov.,666- 667;LF,105.
CRYPTOMELA-NE(monoclinic)	KMn_8O_{16}	$Mn_8{}^o[KO_{16}]^{chh}$	Mon. I 2/m	a=9.956Å β=90.95° b=2.8705Å Z=1 c=9.706Å	(K,□)(2a) $Mn_{I\text{-}II}$(4i) $O_{I\text{-}IV}$(4i)	$Mn_8{}^o[(B,K)O_{16}]^{chh}$ HOLLANDITE	Acta Cryst.,1982,B38,1056- 1065;Pov.,305;SR,13,188-191; LF,107.
CRYPTOMELA-NE(tetragonal)	KMn_8O_{16}	$Mn_8{}^o[KO_{16}]^{chh}$	Tet. I 4/m	a=9.866Å Z=1 c=2.872Å	(K,□)(4e) (occ.0.16) Mn(8h) $O_{I\text{-}II}$(8h)	Dist.deriv. $Mn_8{}^o[(B,K)O_{16}]^{chh}$ HOLLANDITE	Acta Cryist.,1986,B42,162- 167;Pov.,305;Hölzel,73;LF,107.
DANIELSITE	$(Cu,Ag)_{14}HgS_8$		Orth.	a=9.644Å Z=4 b=9.180Å c=18.156Å			Am.Min.,1988,73,187-188; Hölzel,15.
DAVIDITE-(Y)	$Y(Ti,Fe)_{21}O_{38}$	$(Ti,Fe)_{19}{}^o$ $(Ti,Fe)_2{}^t[YO_{38}]^{hhc}$	Trig. $R\bar{3}$	a=10.376Å a_R=9.190Å c=20.910Å α=68.73° Z=3 Z_R=1	$O_{I\text{-}VI}$(6f) O_{VII}(2c) ... (rhomb. descript.)		SR,45A,236;Am.Min.,1979, 64,1010-1017;Pov.,280;Str. Tab.,187.
DIAOYUDAOITE	$NaAl_{11}O_{17}$		Hex. P6/mmc	a=5.602Å Z=2 c=22.626Å			Hölzel,suppl..
ELLISITE	Tl_3AsS_3		Trig. $R\bar{3}m$	a=9.5707Å a_R=5.9967Å c=6.9888Å α=105.88° Z=3 Z_R=1	Tl(3b) As(1a) S(3b)		Zeit.Krist.,1980,249-254; Am.Min.,1979,64,701-707; Hölzel,37.
ENARGITE	Cu_3AsS_4	$Cu_3{}^tAs^t[S_4]^h$ (subs.deriv. Wurtzite-2H)	Orth $Pmn2_1$	a=7.407Å Z=2 b=6.436Å c=6.154Å	As(2a) $S_{I\text{-}II}$(2a) Cu_I(2a) S_{III}(4b) Cu_{II}(4b)	$Cu_3{}^tAs^t[S_4]^h$ ENARGITE	Acta Cryst.,1970,B26,1878- 1897;Pov.,229-230;LF,116; SR,35A,19-20.

Table 60

$AB_mC_n(m,n)\geq3$(cont.)

NAME	CHEMICAL FORMULA	STRUCTURAL FORMULA	SPACE GROUP	UNIT CELL DIMENSIONS	EQUIVALENT POSITIONS	STRUCTURE TYPE	REFERENCES
FAMATINITE	Cu_3SbS_4	$Cu_3{}^tSb^t[S_4]^c$	Tet. $I\bar{4}2m$	a=5.38Å Z=2 c=10.76Å	Sb(2a) S(8i) Cu_I(2b) Cu_{II}(4d)	$Cu_3{}^tSb^t[S_4]^c$ FAMATINITE	SR,21,349-350;Pov.,229-230;LF,112.
FURUTOBEITE	$(Cu,Ag)_6PbS_4$		Mon. Cm...	a=20.055Å β=101.57° b=3.964Å Z=4 c=9.705Å			Bull.Min.,1981,104,737-741;Hölzel,14.
HATRURITE	Ca_3SiO_5		Trig. $R\bar{3}m$	a=7.150Å Z=9 c=25.560 (1000°C)			JCPDS,16-406.
HIBONITE	(Ca,Ce) $(Al,Ti,Mg)_{12}O_{19}$	$(Al,Ti,Mg)_{12}{}^o$ $[(Ca,Ce)O_{19}]^{c/h}$	Hex. $P6_3/mmc$	a=5.61Å c=22.16Å Z=2			RRW,273;Pov.,280-281; Hölzel,65;Str.Tab.,187;N.Jb. Min.Abh.,1968,109,192-200.
HÖGBOMITE-8H	$(Al,Fe,Mg,Ti,Zn)_{11}$ $O_{15}(OH)$	$(Al,Fe,Ti,Zn)_8{}^o$ $(Fe,Zn,Mg,Al)_3{}^t$ $[O_{15}OH]^{ccch}$	Hex. $P6_3mc$	a=5.734Å c=18.389Å Z=2	Al_{I-II}(12d) Al_{III}(2b) ... (part.occ.)		Am.Min.,1982,67,373-380;Pov.,281-282; Hölzel,suppl..
HOLLANDITE	$(Ba,K)Mn_8O_{16}$	$Mn_8{}^o$ $[(Ba,K)O_{16}]^{chh}$	Mon. I 2/m	a=9.91Å β=90°36' b=2.872Å Z=1 c=9.75Å	(Ba,K)(2b) Mn(8h) O_{I-II}(8h) (for I 4/m)	$Mn_8{}^o[(Ba,K)O_{16}]^{chh}$ HOLLANDITE	Wyckoff,3,494-496;Pov., 305-306;SR,13,190-195;LF, 107;Min.Abs.,89M/0251,27.
HYALOPHANE	$(K,Ba)(Al,Si)_4O_8$	$(K,Ba)^{[9]}$ $\{3\infty\}[(Al,Si)_4{}^tO_8]$	Mon. C2/m	a=8.557Å β=115.69° b=13.040Å Z=4 c=7.200Å	(K,Ba)(4i)O_I(4g) $(Al,Si)_{I-II}$(8j) O_{II}(4i) O_{III-V}(8j)		Acta Cryst.,1977,B33,3073-3076;Pov.,741,346;SR,43A, 314-315;LF,267.
KURAMITE	Cu_3SnS_4	$Cu_2{}^tCu^tSn^t[S_4]^c$	Tet. $I\bar{4}2m$	a=5.445Å Z=2 c=10.75Å		$Cu_2{}^tFe^tSn^t(S_4)^c$ STANNITE	Am.Min.,1980,65,1067 (Abs.);Hölzel,19;LF,113.
LABRADORITE	$(Ca,Na)(Si,Al)_4$ O_8	$(Ca,Na)^{[8]}$ $\{3\infty\}[Si_2{}^t(Si,Al)^t$ $Al^tO_8]$	Tric. C1	a=8.1736Å α=93.462° b=12.8736Å β=116.054° c=7.1022Å γ=90.47° Z=4	$(Ca,Na)_{I-II}$(2i) ...	$(Na,Ca)^{[8]}$ $\{3\infty\}[Si_2{}^t(Si,Al)^tAl^tO_8]$ PLAGIOCLASE	Am.Min.,1980,63,81-95; Str.Tab.,478;LF,269.
LINDSLEYITE	(Ba,Sr) $(Ti,Cr,Fe)_{21}O_{38}$		Trig. $R\bar{3}$ ...	a=10.37Å a_R=9.09Å c=20.52Å α=69.6° Z=3 Z_R=1			Am.Min.,1983,68,494-505; Hölzel,64.
LITHIOPHOSPHATE	Li_3PO_4	$Li_3{}^tP^t[O_4]^h$	Orth. Pmnb	a=4.86Å Z=4 b=10.26Å c=6.07Å	Li_I(8d) Li_{II}(4c) P(4c) O_I(8d) O_{II-III}(4c)		Acta Cryst.,1960,13,863-867;Pov.,537;RRW,361.
LITHIOTANTITE	$Li(Ta,Nb)_3O_8$		Mon. $P2_1/c$	a=7.444Å β=107.18° b=5.044Å Z=4 c=15.255Å			Am.Min.,1984,69,1191-1192 (Abs.);Hölzel,78.
LITHIOWODGINITE	$LiTa_3O_8$		Mon. C2/c	a=9.414Å β=91°5' b=11.526Å Z=4 c=5.052Å			Min.Zhurn.,1990,112,94-100;Hölzel,suppl..
LIVINGSTONITE	$HgSb_4S_8$		Mon. A2/a	a=30.567Å β=103.39° b=4.015Å Z=8 c=21.465Å	Hg_I(4b)Hg_{II}(4e) Sb_{I-IV}(8f) S_{I-VIII}(8f)		Zeit.Krist.,1975,141,174-192;SR,41A,17;Pov.,263; SR,21,347-349.
LOVERINGITE	(Ca,Ce,La) $(Ti,Fe,Cr)_{21}O_{38}$	$(Ti,Fe,Cr)_{19}{}^o$ $(Ti,Fe,Cr)_2{}^t$ $[(Ca,Ce,La)O_{38}]^{chh}$	Trig. $R\bar{3}$	a=10.337Å a_R=9.117Å c=20.677Å α=69.07° Z=3 Z_R=1	(Ca,Ce,La)(1a) ...(rhomb.desc.)		Am.Min.,1978,63,28-36; SR,44A,197;Hölzel,65.

Table 61

AB_mC_n(m,n)≥3(cont.)

NAME	CHEMICAL FORMULA	STRUCTURAL FORMULA	SPACE GROUP	UNIT CELL DIMENSIONS	EQUIVALENT POSITIONS	STRUCTURE TYPE	REFERENCES
LUZONITE	Cu_3AsS_4	$Cu_3{}^tAs^t[S_4]^c$	Tet. I $\bar{4}$2m	a=5.290Å Z=2 c=10.465Å	Cu_I(2b) As(2a) Cu_{II}(4d) S(8i)	$Cu_3{}^tSb^t[S_4]^c$ FAMATINITE	SR,21,349-350:Pov.,229-230;LF,112.
MAGNETO-PLUMBITE	$Pb(Fe,Mn)_{12}O_{19}$	$(Fe,Mn)_{12}{}^o[PbO_{19}]^{chhhc}$	Hex. $P6_3/mmc$	a=5.873Å Z=2 c=23.007Å	O_I(4e) ...		Am.Min.,1989,74,1186-1194; Pov.,280-281;RRW,375.
MANGANESE-SHADLUNITE	$(Fe,Cu)_8(Mn,Pb)S_8$	$(Mn,Pb)^o(Fe,Cu)_8{}^t[S_8]^c$	Cub. Fm$\bar{3}$m	a=10.73Å Z=4		$(Ni,Fe)^o(Ni,Fe)_8{}^t[S_8]^c$ PENTLANDITE	ZVMO,1973,102,63-74; Hölzel,16.
MATHIASITE	$(K,Ca,Sr)(Ti,Cr,Fe,Mg)_{21}O_{38}$		Trig. R$\bar{3}$	a=10.35Å a_R=9.119Å c=2.058Å α=69.24° Z=3 Z_R=1	(K,Ca,Sr)(1a) (part.occ.) ... (rhomb.desc.)		Acta Cryst.,1983,C39, 421-422;Am.Min.,1983, 68,494-505;Hölzel,64.
MGRIITE	Cu_3AsSe_3		Cub. Pd$\bar{3}$m	a=5.530Å Z=1			Am.Min.,1983,68,280-281(Abs.);Hölzel,38.
MOROZEVICZI-TE	$Pb_3Ge_{1-x}S_4$		Cub.	a=10.61Å Z=8			Am.Min.,1981,66,437-438(Abs.);Hölzel,18.
NAGYÁGITE	$Au(Pb,Sb,Fe)_8(Te,S)_{11}$		Tet. P$\bar{4}$	a=4.14Å Z=1 c=30.15Å			Am.Min.,1949,34,342-382;Pov.,264;Hölzel,31.
OLIGOCLASE	$(Na,Ca)(Si,Al)_4O_8$	$(Na,Ca)^{[8]}\{3\infty\}[Si_2{}^t(Si,Al)_2{}^tO_8]$	Tric. C1	a=8.1553Å α=93°57.9' b=12.8206Å β=116°28.5' c=7.1397Å γ=88°37.9' Z=4	(Na,Ca)(4i) $(Si,Al)_{I-IV}$(4l) O_{I-VIII}(4i)	$(Na,Ca)^{[8]}$ $\{3\infty\}[Si_2{}^t(Si,Al)^t{}_2O_8]$ OLIGOCLASE	Zeit.Krist.,1971,133,43-65;SR,37A,341;Str.Tab., 477;LF,269.
OLYMPITE	Na_3PO_4		Orth. Pnma...	a=10.154Å Z=16 b=14.819Å c=10.143Å	P_I(4c) P_{II}(8d) ... (Z=8 for $LiNa_5[PO_4]_2$)		Am.Min.,1981,66,438(Abs.); Sov.Phys.Cryst.,1992,37, 772-776;Hölzel,145.
OTJISUMEITE	$PbGe_4O_9$		Tric. P1...	a=6.945Å α=102.94° b=6.958Å β=103.05° c=9.279Å γ=114.77° Z=2			Am.Min.,1987,72,1026-1027(Abs.);Hölzel,85.
PARAPIERRO-TITE	$Tl(Sb,As)_5S_8$		Mon. Pn	a=8.098Å β=91.96° b=19.415Å Z=4 c=9.059Å	Tl_{I-II}(2a) $(Sb,As)_{I-X}$(2a) S_{I-XVI}(2a)		Zeit.Krist.,1980,151,203-216;Am.Min.,1976,61, 504(Abs.);Hölzel,41.
PAVONITE	$(Ag,Cu)(Bi,Pb)_3S_5$		Mon. C2/m	a=13.305Å β=94.02° b=4.042Å Z=4 c=16.417Å	Bi_{I-III}(4i) S_{I-IV}(4i) Ag_I(2a) Ag_{II}(2d)		SR,45A,39-40;Pov.,249; K/S,66-67.
PENZHINITE	$(Ag,Cu)_4Au(S,Se)_4$		Hex. $P6_322$	a=13.779Å Z=18 c=16.980Å			ZVMO,1984,113,356-360; Hölzel,14.
PERMINGEATI-TE	$Cu_3(Sb,As)Se_4$	$Cu_3{}^t(Sb,As)^t[Se_4]^c$ (Subs.d. Sphalerite)	Tet. I $\bar{4}$2m	a=5.63Å c=11.23Å Z=2		$Cu_3{}^tSb^t[S_4]^c$ FAMATINITE	Am.Min.,1972,57,1554-1555 (Abs.);Garcia Orta,Sér. Geol.,1986,9,41-53.
PETRUKITE	$(Cu,Fe,Sn)_3SnS_4$	$(Cu,Fe,Sn)_3{}^tSn^t[S_4]^c$ (Subs.d. Sphalerite)	Orth. $Pmn2_1$	a=7.667Å Z=2 b=6.4399Å c=6.205Å	(Cu,Fe,Sn)(4b) Cu_{II}(2a) S_I(4b) Sn(2a)S_{II-III}(2a)		Can.Min.,1989,27,673-688; Hölzel,suppl..

Table 62

$AB_mC_n(m,n)\geq 3$(cont.)

NAME	CHEMICAL FORMULA	STRUCTURAL FORMULA	SPACE GROUP	UNIT CELL DIMENSIONS	EQUIVALENT POSITIONS	STRUCTURE TYPE	REFERENCES
PLAGIOCLASE	(Na,Ca) $(Si,Al)_4O_8$	$(Na,Ca)^{[8]}\{3\infty\}$ $[Si_2^t(Si,Al)^tAl^tO_8]$	Tric. P1	a=8.152Å α=94.28° b=12.784Å β=116.67° c=7.165Å γ=87.74° Z=4	(Na,Ca)(4i) ...	$(Na,Ca)^{[8]}\{3\infty\}$ $[Si_2^t(Si,Al)^tAl^tO_8]$ PLAGIOCLASE	Am.Min.,1977,62,921-931; Pov.,344-345;Zeit.Krist., 1980,151,77-89;LF,269.
PLUMBOFERRI-TE	$PbFe_4O_7$		Trig. P312	a=11.88Å Z=42 c=47.23Å			JCPDS,22-656;Pov.,280-281;Hölzel,64.
POLKOVICITE	$(Fe,Pb)_3$ $(Ge,Fe)_{1-x}S_4$		Cub.	a=10.61Å Z=8			Am.Min.,1981,66,437-438 (Abs.);Hölzel,18.
PRIDERITE	(K,Ba) $(Ti,Fe)_8O_{16}$	$(Ti,Fe)_8^{[8]}$ $[(K,Ba)O_{16}]^{chh}$	Tet. I 4/m	a=10.139Å Z=1 c=2.9664Å	(Ti,Fe)(8h) (K,Ba)(2b) O_{I-II}(8h)	$Mn_8^{[8]}[(Ba,K)O_{16}]^{chh}$ HOLLANDITE	Acta Cryst.,1982,B38, 1056-1065;Pov.,306.
PROUSTITE	Ag_3AsS_3	$[[Ag]_3$ $\{g\}[As^{[3n]}S_3]]$	Trig. R3c	a=10.82Å a_R=6.89Å c=8.69Å α=103°25' Z=6 Z_R=2	Ag(18b) As(6a) S(18b)	$[[Ag]_3\{g\}[As^{[3n]}S_3]]$ PROUSTITE	SR,31A,19-20;Pov.,239; LF,178;K/S,87-88.
PYRARGYRITE	Ag_3SbS_3	$[[Ag_3]$ $\{g\}[Sb^{[3n]}S_3]]$	Trig. R3c	a=11.04Å a_R=7.01Å c=8.72Å α=104°4' Z=6 Z_R=2	Ag(18b) Sb(6a) S(18b)	$[[Ag]_3\{g\}[As^{[3n]}S_3]]$ PROUSTITE	SR,31A,19-20;Pov.,239; LF,178;K/S,87-88.
PYROSTILPNITE	Ag_3SbS_3		Mon. $P2_1/c$	a=6.84Å β=117°9' b=15.84Å Z=4 c=6.24Å	Sb(4e) Ag_{I-III}(4e) S_{I-III}(4e)		Pov.,239;SR,33A,22-24.
SABATIERITE	Cu_6TlSe_4		Orth.	a=3.986Å Z=1 b=5.624Å c=9.778Å			Bull.Min.,1978,101,557-360;Zeit.Krist.,1987,181, 241-249;Hölzel,15.
SANIDINE	$(K,Na)(Si,Al)_4$ O_8	$(K,Na)^{[9]}$ $\{3\infty\}[(Si,Al)_4{}^tO_8]$	Mon. C2/m	a=8.558Å β=116.07° b=12.997Å Z=4 c=7.179Å	(K,Na)(4i) $(Si,Al)_{I-II}$(8j) O_I(4g) O_{II}(4i) O_{III-V}(8j)	$(K,Na)^{[9]}$ $\{3\infty\}[(Si,Al)_4{}^tO_8]$ SANIDINE	Am.Min.,1978,63,1264127; Pov.,346-347;Am.Min., 1987,72,973-978;LF,266.
SANTANAITE	$Pb_{11}CrO_{16}$		Hex. $P6_322$	a=9.03Å Z=6 c=39.84Å			Am.Min.,1973,58,966-967 (Abs.);Hölzel,139.
SELENOSTE-PHANITE	$Ag_5Sb(Se,S)_4$		Orth. $P2_12_12_1$	a=7.86Å Z=4 b=11.84Å c=8.92Å			Am.Min.,1987,72,225 (Abs.);Hölzel,38.
SENAITE	$Pb(Ti,Fe,Mn)_{21}$ O_{38}	$(Ti,Fe,Mn)_{19}^o$ $(Ti,Fe,Mn)_2^t$ $[PbO_{38}]^{hhc}$	Trig. $R\bar{3}$	a=10.393Å a_R=9.172Å c=20.811Å α=69.02° Z=3 Z_R=1	Pb(1a) $(Ti,Fe,Mn)_1$(1b) ... O_{I-IV}(6f) O_{VI}(2c) ... (rhomb.descript.)		Acta Cryst.,1976,B32,1509-1513;Am.Min.,1968,53,869-879;SR,42A,254;Pov.,280-281.
SHADLUNITE	$(Fe,Cu)_8$ $(Pb,Cd)S_8$	$(Pb,Cd)^o$ $(Fe,Cu)_8{}^t[S_8]^c$	Cub. $Fm\bar{3}m$	a=10.91Å Z=4	(Pd,Cd)(4b) S_I(8c) (Fe,Cu)(32f)S_{II}(24e)	$(Ni,Fe)^o(Ni,Fe)_8{}^t[S_8]^c$ PENTLANDITE	ZVMO,1973,102,63-74; LF,131;Hölzel,16.
SILLÉNITE	$Bi_{12}SiO_{20}$ γ-form		Cub. I 23	a=10.184Å Z=2	Bi(24f)O_I(24f)O_{II}(12d) (Z=12 for Bi_2O_3)		JCPDS,29-235;Hölzel,63; Pov.,271.
SKINNERITE	Cu_3SbS_3		Mon. $P2_1/c$	a=7.81Å β=90°21' b=10.25Å Z=8 c=13.27Å			Am.Min.,1974,59,889-895; Hölzel,37.
STEPHANITE	Ag_5SbS_4		Orth. $Cmc2_1$	a=7.837Å Z=4 b=12.467Å c=8.538Å	Ag_I(4a) S_{I-II}(4a) Ag_{II-III}(8b) S_{III}(8b) Sb(4a)		Acta Cryst.,1970,B26,201-207;SR,35A,15-16;Pov., 240;SR,34A,12-13.

Table 63

$AB_mC_n(m,n)\geq 3$(cont.)

NAME	CHEMICAL FORMULA	STRUCTURAL FORMULA	SPACE GROUP	UNIT CELL DIMENSIONS	EQUIVALENT POSITIONS	STRUCTURE TYPE	REFERENCES
STIBIOCONITE	$Sb_3O_6(OH)$		Cub. Fd3m	a=10.26Å Z=8			Pov.,277;Am.Min.,1952,37,982-989;Hölzel,66.
SULVANITE	Ca_3VS_4	$Ca_3{}^tV^t[S_4]^c$	Cub. P$\bar{4}$3m	a=5.3912Å Z=1	Ca(3d) S(4e) V(1a)		SR,31A,41;Am.Min.,1966,51,890-894;K/S,42-43;Pov.,230.
TAKEUCHIITE	$(Mg,Mn)_3BO_5$		Orth. Pnnm	a=27.585Å Z=24 b=12.561Å c=6.027Å	$(Mg,Mn)_I$(4f) $(Mg,Mn)_{II}$(4e) ...		Zeit.Krist.,1987,181,135-149;Am.Min.,1980,65,1130-1133.
TEMAGAMITE	Pd_3HgTe_3		Orth.	a=11.57Å Z=6 b=12.16Å c=6.76Å			Can.Min.,1973,12,193-198;Hölzel,suppl..
USTARASITE	$Pb(Bi,Sb)_6S_{10}$		?				Am.Min.,1956,41,814-815 (Abs.);Hölzel,44;Pov.,763.
WAKABAYASHILI-TE	$SbAs_{10}S_{18}$		Mon. $P2_1/m$	a=25.17Å β=120° b=6.48Å Z=6 c=25.24Å			Am.Min.,1972,57,1311-1312 (Abs.);RRW,660;Hölzel,46;Can.Min.,1975,13,418-419.
WINSTANLEYITE	$TiTe_3O_8$		Cub. Ia3	a=10.963Å Z=8			Min.Mag.,1979,43,453-457;Hölzel,94.
WITTICHENITE	Cu_3BiS_3		Orth. $P2_12_12_1$	a=7.723Å Z=4 b=10.395Å c=6.716Å	Bi(4a) Cu_{I-III}(4a) S_{I-III}(4a)		Acta Cryst.,1973,B29,2528-2535;SR,39A,29-31;Pov.,254;K/S,81-82.
WITTITE	$Pb_{0.35}Bi_{0.44}(S,Se)$		Mon. P2/m	a=4.19Å β=101.35° b=4.08Å Z=? c=15.56Å			Am.Min.,1980,65,789-796;Hölzel,40;Pov.,765,254.
XANTHOCONITE	Ag_3AsS_3		Mon. C2/c	a=12.00Å β=110°0′ b=6.26Å Z=8 c=17.08Å	As(8f) Ag_{I-III}(8f) S_{I-III}(8f)		Acta Cryst.,1968,B24,77-81;K/S,77-78;Pov.,239;Hölzel,37.
YAFSOANITE	$(Zn,Ca,Pb)_3TeO_6$		Cub. $I4_132$	a=12.621Å Z=16	Te(16e) Zn(24h) Ca(24g) O(96i)		Min.Abs.,85M/0176,17-18
YIMENGITE	$K(Cr,Ti,Fe,Mg)_{12}O_{19}$		Hex. $P6_3/mmc$	a=5.875Å Z=2 c=22.940Å			Am.Min.,1985,70,218(Abs.);Hölzel,84.
YTTROTUNGSTI-TE-(Y)	$(Y,La,Ca)(W,Fe,Si,Al,Ti)_2(O,OH,H_2O)_9$		Mon. $P2_1/m$	a=6.95Å β=104°56′ b=8.64Å Z=2 c=5.77Å			Min.Mag.,1971,38,261-285;Pov.,573;Hölzel,81.

Table 64

$A_pB_qC_r(p,q)\geq2;r\geq3$

NAME	CHEMICAL FORMULA	STRUCTURAL FORMULA	SPACE GROUP	UNIT CELL DIMENSIONS	EQUIVALENT POSITIONS	STRUCTURE TYPE	REFERENCES
AERUGITE	$Ni_{8.5}As_3O_{16}$	$Ni_{8.5}{}^{o}As^{o}As_2{}^{t}[O_{16}]^{c}$	Trig. R$\bar{3}$m	a=5.9511Å a_R=9.8105Å c=27.5676Å α=35,312° Z=3 Z_R=1	Ni_I(3e) As_{II}(2c) Ni_{II}(6h) O_{I-II}(2c)(r.d.) As_I(1a) O_{III-IV}(6h)		Acta Cryst.,1989,B45201-205; Pov.,510;Str.Tab.,318;Min.Mag, 1965,35,72-83;Hölzel,150.
ASCHAMAL-MITE	$Pb_6Bi_2S_9$		Mon. C2/m...	a=13.71Å β=91.0° b=4.09Å Z=4 c=31.43Å			Am.Min.,1984,69,810-815; Hölzel,44.
AUGITE	$(Ca,Mg,Fe)_2(Si,Al)_2O_6$	$Ca^{[8]}(Mg,Fe)^{o}$ $\{1\infty\}[(Si,Al)_2{}^{t}O_6]$	Mon. C2/c	a=9.750Å β=74°10' b=8.930Å Z=4 c=5.299Å		$Ca^{[8]}Mg^{o}$ $\{1\infty\}[Si_2{}^{t}O_6]$ DIOPSIDE	Wyckoff,4,295;Pov.,424;Str. Tab.,413;SR,39A,336,41A,394; Hölzel,212;LF,205.
BARIOPYRO-CHLORE	$(Ba,Sr)_2(Nb,Ti)_2(O,OH)_7$	$(Ba,Sr)_2{}^{cb}(Nb,Ti)_2{}^{[6]}[(O,OH)_7\square]^{Cs}$	Cub. Fd$\bar{3}$m	a=10.562Å Z=8			Min.Mag,1959,32,10-25;Am. Min.,1977,62,403-410;Hölzel,66; LF,140.
BEUSITE	$(Mn,Fe,Ca,Mg)_3(PO_4)_2$	$(Mn,Fe,Ca,Mg)^{[7]}$ $(Mn,Fe,Ca,Mg)_2{}^{[5]}$ $\{g\}[P^{t}O_4]_2$	Mon. $P2_1/c$	a=8.78Å β=99°25' b=11.52Å Z=4 c=6.15Å			Am.Min.,1968,53,1799-1814; RRW,68;Pov.,537;Str.Tab.,309; Hölzel,142.
BISMUTO-HAUCHE-CORNITE	$Ni_9Bi_2S_8$		Tet. P4/mmm	a=7.300Å Z=1 c=5.402Å			Min.Mag.,1980,43,873-876.
BLOSSITE	$Cu_2V_2O_7$ α-form	$Cu_2{}^{[5]}\{g\}[V_2{}^{t}O_7]$	Orth. Fdd2	a=20.676Å Z=8 b=8.392Å c=6.446Å	(for synthetic m.) Cu(16b) O_I(8a) V(16b) O_{II-IV}(16b)		Am.Min.,1987,72,397-400;Acta Cryst.,1975,B31,603-605;Hölzel, 143.
BOGDANOVI-TE	$Au_5(Cu,Fe)_3(Te,Pb)_2$		Cub.	a=4.087Å Z=?			Am.Min.,1979,64,1329(Abs.); Hölzel,1.
BOULANGE-RITE	$Pb_5Sb_4S_{11}$	$Pb_5{}^{[6/7]}Sb_4{}^{[6/7]}[S_{11}]^{\#}$	Orth. Pbnm	a=21.30Å Z=4 b=23.47Å c=4.04Å	Pb_{I-III}(4c) $(Pb,Sb)_{I-III}$(4c) ...	$Pb_5{}^{[6/7]}Sb_4{}^{[6/7]}$ $[S_{11}]^{\#}$ BOULANGERI-TE	Min.Abs.,80/0161,20;Am.Min., 1960,45,1266-1271;LF,164; K/S,68-69.
BROWNMI-LLERITE	$Ca_2(Al,Fe)_2O_5$		Orth. Ibm2	a=5.584Å Z=4 b=14.60Å c=5.374Å	Ca(8c) O_{I-II}(8c) $(Fe,Al)_I$(4a) O_{III}(4b) $(Al,Fe)_{II}$(4b)		Acta Cryst.,1971,B27,2311-2315;Pov.,289;Am.Min.,1965, 50,2106(Abs.);SR,37A,259.
BURSAITE	$Pb_5Bi_4S_{11}$		Orth. Bbmm	a=13.42Å Z=2 b=20.37Å c=4.12Å			N.Jb.Min.Abh.,1988,158,293-309;Hölzel,44.
CALCIOBE-TAFITE	$(Ca,Na)_2(Nb,Ti)_2(O,OH)_7$	$(Ca,Nb)_2{}^{cb}$ $(Nb,Ti)_2{}^{[6]}$ $[(O,OH)_7\square]^{Cs}$	Cub. Fd$\bar{3}$m	a=10.2978Å Z=8	(Ca,Na)(16c) (Nb,Ti)(16c) (O,OH)(8b) O(48f)		Am.Min.,1983,68,262-276;LF, 140.
CANNIZZARI-TE	$Pb_{46}Bi_{54}S_{127}$	$Pb_{46}{}^{o}Bi_{54}{}^{o}[S_{127}]^{\#}$	Mon. $P2_1/m$	a=189.8Å β=11.93° b=4.09Å Z=2 c=74.06Å	Pb(...) ...		Acta Cryst.,1979,B35,133-136; Pov.,250-251;Hölzel,44.
CHEKHOVI-CITE	$Bi_2Te_4O_{11}$		Mon. $P2_1$	a=19.00Å β=95.67° b=7.982Å Z=4 c=6.938Å			Am.Min.,1989,74,1400(Abs.); Hölzel,94.

Table 65

$A_pB_qC_r(p,q)\geq2;r\geq3$(cont.)

NAME	CHEMICAL FORMULA	STRUCTURAL FORMULA	SPACE GROUP	UNIT CELL DIMENSIONS	EQUIVALENT POSITIONS	STRUCTURE TYPE	REFERENCES
CHERVETITE	$Pb_2V_2O_7$	$Pb_2^{[8/9]}\{g\}[V_2^tO_7]$	Mon. $P2_1/a$	a=13.37Å β=106° b=7.16Å Z=4 c=7.11Å	$Pb_{I-II}(4e)$ $V_{I-II}(4e)$ $O_{I-VII}(4e)$		Bull.Min.,1967,74,279-284; SR,32A,275-276;Pov.,498; Am.Min.,1963,48,1416(Abs.).
CHIOLITE	$Na_5Al_3F_{14}$		Tet. P4/mnc	a=7.00Å Z=2 c=10.39Å	$Al_I(2a)$ $Na_I(2b)$ $Al_{II}(4c)$ ...		SB,6,31-32,121-122;Pov., 668;RRW,125;Hölzel,51.
CLARKEITE	$(Na,Ca,Pb)_2$ $U_2(O,OH)_7$		?				Am.Min.,1956,41,127-133; Hölzel,89;Str.Tab.,226;Pov., 333.
COMANCHEITE	$Hg_{13}O_9$ $(Cl,Br)_8$		Orth. Pnnm...	a=18.41Å Z=4 b=21.64Å c=6.677Å			Can.Min.,1981,19,393-396; Hölzel,55.
CORDEROITE	$Hg_3S_2Cl_2$ α-form		Cub. $I\,2_13$	a=8.94Å Z=4			Am.Min.,1974,59,652-655; Hölzel,55
COSALITE	$Pb_2Bi_2S_5$	$Pb_2^oBi_2^{[5/6]}[S_5]$ #	Orth. Pbnm	a=19.098Å Z=8 b=23.890Å c=4.057Å	$Pb_{I-IV}(4c)$ ...	$Pb_2^oBi^{[5/6]}[S_5]$ # COSALITE	Zeit.Krist.,1974,140,114-136; SR,40A,37-38;K/S,64-65;Hölzel, 37;LF,162;Wyckoff,3,317-318.
CUPROBISMUTITE	$Cu_4Bi_7S_{12}$	$Cu_4^{t/tr}Bi_7^o[S_{12}]$ #	Mon. C2/m	a=17.52Å β=101.26° b=3.912Å Z=6? c=15.24Å	$Bi_I(2c)$ ...	$Cu_4^{t/tr}Bi_2^o[S_{12}]$ # CUPROBISMU- TITE	Acta Cryst.,1975,B31,703- 708:Am.Min.,1952,37,447- 452;K/S,60-61;Pov.,255-256.
DIOMIGNITE	$Li_2B_4O_7$		Tet. $I\,4_1cd$	a=9.470Å Z=8 c=10.279Å	Li(16b) $O_{I-III}(16b)$ $B_{I-II}(16b)$ $O_{IV}(8a)$		Can.Min.,1987,25,173-180; Acta Cryst.,1962,15,190-193.
DOLORESITE	$V_3O_4(OH)_4$	$2\infty[V_3^oO_4(OH)_4]$	Mon. C2/m	a=19.64Å β=103°5' b=2.99Å Z=2 c=4.83Å	$V_I(2c)$ $V_{II}(4i)$ $O_{I-IV}(4i)$		Acta Cryst.,1958,11,56-58; Am.Min.,1960,45,1144-1173; Pov.,321-322.
DONHARRISITE	$Ni_9Hg_3S_9$		Mon. C2/m...	a=11.66Å β=97.43° b=6.91Å Z=2 c=10.92Å			Can.Min.,1989,27,257-262; Hölzel,suppl..
DUFRÉNOYSITE	$Pb_2As_2S_5$		Mon. $P2_1$	a=7.90Å β=90°21' b=25.74Å Z=8 c=8.37Å	$Pb_{I-VIII}(2a)$ $As_{I-VIII}(2a)$ $S_{I-XX}(2a)$		Zeit.Krist.,1969,130,15-40;1967, 124,409-419;SR,32A,227-228; Pov.,235;K/S,73-74;Hölzel,42.
EDGARBAILEYITE	$Hg_6Si_2O_7$		Mon. C2/m	a=11.725Å β=112.07° b=7.698Å Z=2 c=5.967Å			Min.Rec,,1950,8,215-220; Hölzel,suppl;Am.Min.,1990,75, 1192-1196.
ERTIXIITE	$Na_2Si_4O_9$		Cub.	a=5.975Å Z=1			Min.Abs.,86M/2251,239; Hölzel,226.
EULYTITE	$Bi_4(SiO_4)_3$	$3\infty[Bi_4^{[3n]}(Si^tO_4)_3]$	Cub. $I\,\bar{4}3d$	a=10.300Å Z=4	Bi(16c) Si(12a) O(48e)	$3\infty[Bi_4^{[3n]}(Si^tO_4)_3]$ EULYTITE	Zeit.Krist.,1966,123,73-76; Pov.,341-342.
FARRINGTONITE	$Mg_3(PO_4)_2$		Mon. $P2_1/n$	a=7.548Å β=95.12° b=8.490Å Z=2 c=5.050Å	$Mg_I(4e)$ $Mg_{II}(2b)$ P(4e) $O_{I-IV}(4e)$		Am.Min.,1985,70,624-629: Am.Min.,1973,58,949-951: Hölzel,143.

Table 66

ApBqCr(p,q)≥2;r≥3(cont.)

NAME	CHEMICAL FORMULA	STRUCTURAL FORMULA	SPACE GROUP	UNIT CELL DIMENSIONS	EQUIVALENT POSITIONS	STRUCTURE TYPE	REFERENCES
FERROSILITE	$(Fe,Mg)_2(SiO_3)_2$	$(Fe,Mg)^{[6]}(Fe,Mg)^{o}$ $\{1\infty\}[Si_2{}^{t}O_6]$	Orth. Pbca	a=18.418Å Z=8 b=9.078Å (24°C) c=5.2366Å	$(Fe,Mg)^{[6]}(8c)$ $(Fe,Mg)^{o}(8c)$ $Si_{I-II}(8c)$ $O_{I-VI}(8c)$	$Mg^{[6]}Mg^{o}\{1\infty\}[Si_2{}^{t}O_6]$ ENSTATITE	Am.Min.,1976,61,38-53; Pov.,737,408;Hölzel,suppl.; LF,207;
FIEDLERITE	$Pb_3Cl_4(OH)_2$		Mon. $P2_1/a$	a=16.62Å β=102°12' b=8.02Å Z=4 c=7.20Å			Pov.,654-655;Hölzel,56.
FREUDEN-BERGITE	$Na_2(Ti,Fe)_8O_{16}$		Mon. C2/m	a=12.267Å β=107.16° b=3.823Å Z=1 c=6.483Å	Na(2a) $(Ti,Fe)_{I-II}(4i)$ $O_{I-IV}(4i)$		Acta Cryst.,1978,B34, 255256;Hölzel,65;Pov., 280-281;Str.Tab.,193.
FÜLÖPPITE	$Pb_3Sb_8S_{15}$		Mon. C2/c	a=13.441Å β=94.71° b=11.726Å Z=4 c=16.930Å	$Pb_I(4e)$ $S_I(8f)$ $Pb_{II}(8f)$ $S_{II}(4e)$ $Sb_{I-IV}(8f)$ $S_{III-VIII}(8f)$		Acta Cryst.,1975,B31,151- 157;SR,41A,1415;Pov.,234; K/S,79-81.
GEOCRONI-TE	$Pb_{14}(Sb,As)_6S_{23}$		Mon. $P2_1/m$	a=8.963Å β=118.02° b=31.93Å Z=2 c=8.500Å	$Pb_{I-VI}(4f)$ $Pb_{VII-VIII}(2e)$ ...		Am.Min.,1976,61,963-970; SR,43A,13;K/S,73.
GIRAUDITE	$(Cu,Zn,Ag)_{12}$ $(As,Sb)_4(Se,S)_{13}$	$3\infty[(Cu,Zn,Ag)_6{}^{t}$ $(Cu,Zn,Ag)_6{}^{tr}$ $(As,Sb)_4{}^{[3n]}(Se,S)_{13}]$	Cub. $I\bar{4}3m$	a=10.578Å Z=2		$3\infty[Cu_6{}^{t}Cu_6{}^{tr}Sb_4{}^{[3n]}S_{13}]$ TETRAHEDRITE	Am.Min.,1982,57,1074-1082 (Abs.);Hölzel,21;LF,283.
GOLDFIELDI-TE	$Cu_{12}(Te,Sb,As)_4$ S_{13}	$3\infty[Cu_6{}^{t}Cu_6{}^{tr}$ $(Te,Sb,As)_4{}^{[3n]}S_{13}]$	Cub. $I\bar{4}3m$	a=10.304Å Z=2		$3\infty[Cu_6{}^{t}Cu_6{}^{tr}Sb_4{}^{[3n]}S_{13}]$ TETRAHEDRITE	JCPDS,29-531;Pov.,739,233- 234;Zeit.Krist.,1988,185,600.
GORTDRUMI-TE	$(Cu,Fe)_6Hg_2S_5$		Orth.	a=14.96Å Z=12 ? b=7.90Å c=24.Å			Min.Mag.,1983,47,35-36; Hölzel,15.
GRAFTONITE	$(Fe,Mn,Ca)_3$ $(PO_4)_2$	$(Fe,Mn,Ca)^{[7]}$ $(Fe,Mn,Ca)_2{}^{[5]}$ $\{g\}$ $[P^{t}O_4]_2$	Mon. $P2_1/c$	a=8.91Å β=98.9° b=11.58Å Z=4 c=6.239Å	$(Fe,Mn,Ca)_{I-III}(4e)$ $P_{I-II}(4e)$ $O_{I-VIII}(4e)$		Am.Min.,1968,53,742-750; Am.Min.,1982,67,826-832; SR,33A,401;Pov.,537.
GRATONITE	$Pb_9As_4S_{15}$		Trig. R3m	a=17.758Å a_R=10.58Å c=7.807Å α=114°10' Z=3 Z_R=1	$Pb_I(18c)$ $As_{II}(3a)$ $Pb_{II}(9b)$ $S_I(18c)$ $As_I(9b)$ $S_{II-IV}(9b)$		Zeit.Krist.,1969,128,321- 338;Pov.,236-237;K/S,70-71.
HÄGGITE	$V_2O_2(OH)_3$		Mon. C2/m	a=12.17Å β=98°15' b=2.99Å Z=2 c=4.83Å	V(4i) $O_I(2c)$ $O_{II-III}(4i)$		Am.Min.,1960,45,1144- 1166;Acta Cryst.,1958,11, 56-58;Pov.,328,144-145.
HAKITE	$(Cu,Hg,Ag)_{12}$ $Sb_4(Se,S)_{13}$	$3\infty[(Cu,Hg,Ag)_6{}^{t}$ $(Cu,Hg,Ag)_6{}^{tr}$ $Sb_4{}^{[3n]}(Se,S)_{13}]$	Cub. $I\bar{4}3m$	a=10.83Å Z=2		$3\infty[Cu_6{}^{t}Cu_6{}^{tr}Sb_4{}^{[3n]}S_{13}]$ TETRAHEDRITE	Am.Min.,1972,57,1553- 1554(Abs.).
HAYCOCKITE	$Cu_4Fe_5S_8$	$Cu_4{}^{t}Fe_4{}^{t}Fe^{o}[S_8]^{c}$	Orth. P222 ...	a=10.705Å Z=12 b=10.734Å c=31.630Å	$Cu_I(2i)$ $Cu_{II}(2k)$ $Cu_{III}(2j)$ $Cu_{IV}(2l)$ $Cu_{V-VIII}(4u)$ ...		Acta Cryst.,1975,B31,2105- 2112;G.O.,Sér.Geol.,1986,9,41- 53;Am.Min.,1972,57,689-708.
HENRYITE	$Cu_4Ag_3Te_4$	$Cu_4{}^{t/tr}Ag_3{}^{o}[Te_4]^{c}$	Cub.	a=12.20Å Z=8			Bull.Min.,1983,106,511-517; Hölzel,14.

Table 67

ApBqCr(p,q)≥2;r≥3(cont.)

NAME	CHEMICAL FORMULA	STRUCTURAL FORMULA	SPACE GROUP	UNIT CELL DIMENSIONS	EQUIVALENT POSITIONS	STRUCTURE TYPE	REFERENCES
HETEROMORPHITE	$Pb_7Sb_8S_{19}$		Mon. C2/c	a=13.628Å β=90°55' b=11.943Å Z=4 c=21.285Å	$Pb_I(4e)$ $S_I(4e)$ $Pb_{II-IV}(8f)$ $S_{II-X}(8f)$ $Sb_{I-IV}(8f)$		Zeit.Krist.,1980,151,193-202; Pov.,234;Hölzel,43.
HODRUSHITE	$Cu_4Bi_6S_{11}$		Mon. A2/m	a=27.21Å β=92°9' b=3.93Å Z=4 c=17.58Å			Min.Mag.,1970,37,641-648; Hölzel,37;Pov.,254;Acta Cryst.,1975,B31,703-708.
HUTCHINSONITE	$(Tl,Pb)_2As_5S_9$		Orth. Pbca	a=10.81Å Z=8 b=35.36Å c=8.16Å	$(Tl,Pb)_{I-II}(8c)$ $As_{I-V}(8c)$ $S_{I-IX}(8c)$		Zeit.Krist.,1965,121,321-348; SR,30A,355-357;K/S,75-77; Pov.,236.
HYDROROMARCHITE	$Sn_3O_2(OH)_2$		Tet. P4mnc ...	a=7.98Å c=9.17Å Z=4			Am.Min.,1973,58,552; Hölzel,58.
IMHOFITE	$Tl_6As_{15}S_{25}$		Mon. $P2_1/n$	a=8.755Å β=108.28° b=24.425Å Z=1 c=5.739Å	$Tl_{I-III}(4e)$ $As_{I-IV}(4e)$ $S_{I-VII}(4e)$ (part. occ.)		Zeit.Krist.,1976,144,323-333: Pov.,236,K/S,76-77;SR,43A,24; Am.Min.,1966,51,531-532(Abs.).
IRTYSHITE	$Na_2(Ta,Nb)_4O_{11}$		Hex. $P6_3/m$...	a=6.231Å Z=6 c=36.77Å			Am.Min.,1986,71,1545(Abs.); Hölzel,77.
JANGGUNITE	$(Mn,Fe)_6O_8(OH)_6$		Orth.	a=9.324Å Z=4 b=14.05Å c=7.956Å			Min.Mag.,1977,41,519-523; Hölzel,74.
JEPPEITE	$(K,Ba)_2(Ti,Fe)_6O_{13}$		Mon. C2/m	a=15.453Å β=99.25° b=3.8368Å Z=2 c=9.123Å			Min.Mag.,1984,48,263-266; Am.Min.,1978,63,795(Abs.); Hölzel,73.
JIMBOITE	$Mn_3(BO_3)_2$	$Mn_3{}^{o}B_2{}^{tr}[O_6]^{n}$ (=Kotoite)	Orth. Pnmm	a=5.658Å Z=2 b=8.740Å c=4.646Å	$Mn_I(2a)$ $O_I(4g)$ $Mn_{II}(4f)$ $O_{II}(8h)$ B(4g)		Sov.Phys.Cryst.,1979,23,272-273;Pov.,466;Am.Min.,1963,48,1416-1417(Abs.).
JORDANITE	$Pb_{14}(As,Sb)_6S_{23}$	$(Pb_{14},(As,Sb)_6,\Box_3)^{o}[S_{23}]^{c}$	Mon. $P2_1/m$	a=8.918Å β=117.79° b=31.899Å Z=2 c=8.462Å	$Pb_{I-VI}(4f)$ $As_{I-II}(4f)$ $Pb_{VII-VIII}(2e)$ $As_{III-IV}(2e)$ ...	Dist.def.deriv. $Na^{o}[Cl]^{c}$ HALITE	Zeit.Krist.,1974,139,161-185: SR,40A,19-20;SR,31A,164;Pov.,237;K/S,72-73.
KALIPYROCHLORE	$(K,H_2O)_2Nb_2(O,OH)_7$	$(K,H_2O)_2{}^{cb}Nb_2{}^{[6]}[(O,OH)_7\Box]^{Qs}$	Cub. $Fd\bar{3}m$	a=10.56Å Z=8			Am.Min.,1978,63,528-530: Hölzel,66.
KAMIOKITE	$Fe_2Mo_3O_8$	$Fe^{o}Fe^{t}Mo_3{}^{o}[O_8]^{ch}$	Hex, $P6_3mc$	a=5.781Å c=10.060Å Z=2	$Fe_{I-II}(2b)$ $O_I(2a)$ Mo(6c) $O_{II}(2b)$ $O_{III-IV}(6c)$		Acta Cryst.,1986,C42,9-11; Hölzel,75.
KARIBIBITE	$Fe_2As_4(O,OH)_9$		Orth.	a=27.91Å Z=6 b=6.53Å c=7.20Å			Am.Min.,1974,59,382(Abs.); Hölzel,91.
KEIVIITE-(Y)	$(Y,Yb)_2Si_2O_7$		Mon. C2/m	a=6.845Å β=101.65° b=8.960Å Z=2 c=4.734Å			Am.Min.,1988,73,191-192(Abs,); Hölzel,197.
KEIVIITE-(Yb)	$(Yb,Y)_2Si_2O_7$		Mon. C2/m	a=6.840Å β=102.11° b=8.916Å Z=2 c=4.745Å			Am.Min.,1984,69,1191(Abs.); Hölzel,197.

Table 68

ApBqCr(p,q)≥2;r≥3(cont.)

NAME	CHEMICAL FORMULA	STRUCTURAL FORMULA	SPACE GROUP	UNIT CELL DIMENSIONS	EQUIVALENT POSITIONS	STRUCTURE TYPE	REFERENCES
KEYITE	$(Cu,Zn,Cd)_3(AsO_4)_2$		Mon. I 2/m...	a=11.65Å β=°98.95° b=12.68Å Z=6 c=6.87Å			Min.Rec.,1977,8,87-90; Hölzel,143.
KILCHOANITE	$Ca_3Si_2O_7$		Orth. Imam...	a=11.42Å Z=8 b=5.09Å c=21.95Å	Ca_I(4a) Si_I(4b) Ca_{II}(4b) Si_{II}(4c) Ca_{III-IV}(8c) ...		Min.Mag.,1971,38,26-31;Am. Min.,1961,46,1203(Abs.);SR, 37A,336;Pov.,396-397.
KOTOITE	$Mg_3(BO_3)_2$	$Mg_3{}^oB_2{}^{tr}[O_6]^h$	Orth. Pnmn	a=5.398Å Z=2 b=8.416Å c=4.497Å	Mg_I(2a) O_I(4g) Mg_{II}(4f) O_{II}(8h) B(4g)		SR,11,425-427;Pov.,464,466; JCPDS,5-648.
KUTINAITE	$Ag_6Cu_{14}As_7$		Cub. I $\bar{4}$3d	a=11.76Å Z=4 ?			Pov.,205;K/S,168-169;Hölzel, suppl..
KYZYLKUMITE	$V_2Ti_3O_9$		Mon.	a=33.80Å β=93.40° b=4.578Å Z=18 c=19.99Å			Am.Min.,1982,67,855-856 (Abs.);Hölzel,73.
LAMMERITE	$Cu_3[(As,P)O_4]_2$	$Cu_3{}^o(As,P)_2{}^t[O_8]^c$	Mon. $P2_1/a$	a=5.079Å β=111.72° b=11.611Å Z=2 c=5.394Å	Cu_I(2d) O_{I-IV}(4e) Cu_{II}(4e) (As,P)(4e)		Am.Min.,1986,71,206-209; Am.Min.,1982,67,415(Abs.); K/B,144-145;Hölzel,143.
LAROSITE	$(Cu,Ag)_{21}(Pb,Bi)_2S_{13}$		Orth.	a=22,15Å Z=10 b=24.03Å c=11.67Å			Can.Min.,1972,11,886-891; Hölzel,40.
LAUNAYITE	$Pb_{22}(Sb,As)_{26}S_{61}$		Mon. C2...	a=42.60Å β=102°5' b=8.04Å Z=4 c=32.3Å			Can.Min.,1967,9,191-213;Pov., 253;Str.Tab.,149;Am.Min.,1968, 53,1423(Abs.);Hölzel,44.
LAVRENTIEVITE	$Hg_3S_2(Cl,Br)_2$		Mon. P2/m...	a=8.94Å β=92.44° b=5.194Å Z=5 c=18.33Å			Am.Min.,1985,70,873-874 (Abs.);Hölzel,55.
LILLIANITE	$Pb_3Bi_2S_6$	$Pb_3{}^oBi_2{}^o[S_6]$ *	Orth. Bbmm	a=13.535Å Z=4 b=20.451Å c=4.104Å	Pb_I(4e) $(Pb,Bi)_{I-II}$(8f) S_I(8f) ...		Acta Cryst.,1972,B28,649-651;SR,38A,36;Pov.,252; Am.Min.,1969,54,579(Abs.).
LIVEINGITE	$Pb_9Ag_{13}S_{28}$	$Pb_9{}^o[As_{13}S_{28}]$ *	Mon. $P2_1$	a=8.44Å β=90° b=69.11Å Z=4 c=7.929Å			Am.Min.,1969,54,1498(Abs.); Pov.,235;Str.Tab.,146.
LOPEZITE	$K_2Cr_2O_7$	$K_2{}^{[6/8]}\{g\}[Cr_2{}^tO_7]$	Tric. P1	a=7.52Å α=98°0' b=13.43Å β=90°51' c=7.39Å γ=96°13' Z=4			Pov.,575-576;Str.Tab.,301; Hölzel,139.
MADOCITE	$Pb_{17}(Sb,As)_{16}S_{41}$		Orth. Pba2...	a=27.2Å Z=4 b=34.1Å c=8.12Å			Can.Min.,1967,9,7-24;Pov., 253:Hölzel,45;RRW,371.
MAYENITE	$Ca_{12}Al_{14}O_{33}$		Cub. I $\bar{4}$3d	a=12.024Å Z=3			Am.Min.,1965,50,2106-2107; Str.Tab.,176;Pov.,289.
MCBIRNEYITE	$Cu_3(VO_4)_2$		Tric. P¯1	a=5.3418Å α=88. 61° b=6.5100Å β=68.11° c=5.1798Å γ=69.22° Z=1			Am.Min.,1988,73,1495(Abs.); Hölzel,143.

Table 69

$A_pB_qC_r(p,q)\geq 2;r\geq 3$(cont.)

NAME	CHEMICAL FORMULA	STRUCTURAL FORMULA	SPACE GROUP	UNIT CELL DIMENSIONS	EQUIVALENT POSITIONS	STRUCTURE TYPE	REFERENCES
MENDIPITE	$Pb_3O_2Cl_2$		Orth. $P2_12_12_1$	a=9.52Å Z=4 b=11.95Å c=5.87Å	$Pb_{I-III}(4c)$ $Cl_{I-II}(4c)$ O(8d) (for Pnma)		RRW,391-392;Str.Tab.,169; Bull.Min.,1971,94,323-331;Pov., 651-652;Hölzel,56;SR,37A,209..
MILLOSEVI-CHITE	$(Al,Fe)_2(SO_4)_3$		Trig.				NBS Monogr.,1978,15,8- ; Hölzel,120;Am.Min.,1974,59, 1140-1141(Abs.).
MOOIHOEKI-TE	$Cu_9Fe_9S_{16}$	$Cu_9{}^tFe_9{}^t[S_{16}]^c$	Tet. $P\bar{4}2m$	a=10.585Å c=5.383Å Z=1	$Cu_I(4n)$ $Cu_{III}(1b)$ $Cu_{II}(4i)$ $Fe_I(4l)$ ...		Acta Cryst.,1973,B29,2365-2372;Am.Min.,1972,57,689-708; SR,39A,52-53;Hölzel,18.
NATROSILITE	$Na_2Si_2O_5$		Mon. $P2_1/a$	a=12.30Å β=104°14' b=4.88Å Z=4 c=8.27Å			Am.Min.,1976,61,339-340 (Abs.);Hölzel,226.
NATROTANTI-TE	$Na_2Ta_4O_{11}$		Trig. $R\bar{3}c$	a=6.2092Å c=36.619Å Z=6	Na(12c) $O_I(36f)$ $Ta_I(18e)$ $O_{II}(18e)$ $Ta_{II}(6b)$ $O_{III}(12c)$		Bull.Min.,1985,108,541-549; Am.Min.,1982,67,413(Abs.); Hölzel,79.
NOLANITE	$(V,Fe,Ti)_{10}$ $O_{14}(OH)_2$	$(V,Fe,Ti)_8{}^o$ $(V,Fe,Ti)_2{}^t$ $[O_{14}(OH)_2]^{ch}$	Hex. $P6_3mc$	a=5.890Å c=9.255Å Z=1	$(V,Fe,Ti)_{I-II}(2b)$ $(V,Fe,Ti)_{III}(6c)$ $O_I(2a)$ $O_{II}(2b)$ $O_{III-IV}(6c)$		Am.Min.,1983,68,833-839; Acta Cryst.,1958,11,703-709; Pov.,282;SR,22,436.
NUKUNDAMI-TE	$Cu_{3.4}Fe_{0.6}S_4$	$(Cu_{3.4},Fe_{0.6})^t$ $[S_4]^c$	Trig. $P\bar{3}m1$	a=3.7830Å c=11.1950Å Z=1	$(Cu,Fe)_{I-II}(2d)$ $S_I(2c)$ $S_{II}(2d)$	Dist.deriv. $Zn^t[S]^c$ SPHALERITE	Am.Min.,1961,66,398-402; Hölzel,26;Min.Mag.,1979,43, 193-200.
ONORATOITE	$Sb_8O_{11}Cl_2$		Mon. C2/m	a=19.047Å β=110.25° b=4.0530Å Z=2 c=10.318Å	$Sb_I(4i)$ $Sb_{II}(4h)$ ...		Acta Cryst.,1984,C40,1506-1510;Min.Mag.,1968,36,1037-1044;Pov.,654-655;Hölzel,57.
PARKERITE	$Ni_3(Bi,Pb)_2S_2$		Orth. Pmam	a=5.545Å Z=1 b=5.731Å c=4.052Å	$Ni_I(2b)$ $Ni_{II}(2e)$(occ.0.5) Bi(2e) S(2f)		Am.Min.,1973,58,435-439:Pov., 224-225;Am.Min.,1974,59,296-301;Hölzel,16;Str.Tab.,112.
PAULMOOREI-TE	$Pb_2As_2O_5$		Mon. $P2_1/a$	a=13.584Å β=108.78° b=5.650Å Z=4 c=8.551Å	$Pb_{I-II}(4e)$ $As_{I-II}(4e)$ $O_{I-V}(4e)$		Am.Min.,1980,65,340-345;Am. Min.,1979,64,352-354;SR,46A, 338-339;Hölzel,91.
PEARCEITE	$(Ag,Cu)_{16}As_2S_{11}$		Mon. C2/m	a=12.59Å β=90.0° b=7.27Å Z=2 c=11.81Å			Am.Min.,1963,48,565-572; Pov.,240;Str.Tab.,141;Hölzel,38.
PIERROTITE	$Tl_2(Sb,As)_{10}S_{16}$		Orth. $Pna2_1$	a=38.746Å Z=4 b=8.816Å c=7.989Å	$Tl_{I-II}(4a)$ $S_{I-XVI}(4a)$ $Sb_{I-VI}(4a)$ $As_{I-IV}(4a)$		Zeit.Krist.,1983,165,209-215; Am.Min.,1972,57,1909-1910 (Abs.);Hölzel,41;Pov.,236.
PINCHITE	$Hg_5O_4Cl_2$		Orth. Ibam	a=11.6Å Z=4 b=6.07Å c=11.7Å			Can.Min.,1974,12,417-418; Hölzel,55.
PLAGIONITE	$Pb_5Sb_8S_{17}$	$Pb_4{}^oPb^{[8]}Sb_6{}^{sq}$ $Sb_2{}^{tr}[S_{17}]^{\#}$	Mon. C2/c	a=13.4857Åβ=107.168° b=11.8656Å Z=4 c=19.9834Å	$Pb_I(4e)$ $S_I(4e)$ $Pb_{II-III}(8f)$ $S_{II-IX}(8f)$ $Sb_{I-IV}(8f)$	$Pb_4{}^oPb^{[8]}Sb_6{}^{sq}$ $Sb_2{}^{tr}[S_{17}]^{\#}$ PLAGIONITE	Zeit.Krist.,1974,139,351-378; K/S,79-80;Pov.,234;Min.Abs., 71-1809.

Table 70

$A_pB_qC_r(p,q)\geq2;r\geq3$(cont.)

NAME	CHEMICAL FORMULA	STRUCTURAL FORMULA	SPACE GROUP	UNIT CELL DIMENSIONS	EQUIVALENT POSITIONS	STRUCTURE TYPE	REFERENCES
PLAYFAIRITE	$Pb_{16}(Sb,As)_{18}S_{43}$		Mon. $P2_1/m$ ...	a=45.4Å β=92°30' b=8.29Å Z=4 c=21.3Å			Can.Min.,1967,9,191-213; Pov.,353;RRW,484;Hölzel,44.
POLYBASITE	$(Ag,Cu)_{16}Sb_2S_{11}$		Mon. C2/m	a=26.17Å β=90.0° b=15.11Å Z=16 c=23.94Å			Am.Min.,1963,48,565-572; Pov.,240;Hölzel,38;Str.Tab., 141.
PSEUDORUTILE	$Fe_2Ti_3O_9$	$Fe_2{}^{o}Ti_3{}^{o}[O_9]^{n}$	Hex. $P6_322$	a=14.375Å Z=5 c=4.615Å	$(Fe,Ti)_I(2a)$ $(Fe,Ti)_{II-V}(6g)$ ... (part.occ.)		Am.Min.,1975,60,898-906; SR,41A,232;Hölzel,68.
RANKAMAITE	$(Na,K,Pb,Li)_3(Ta,Nb,Al)_{11}(O,OH)_{30}$	$(Ta,Nb,Al)_{11}{}^{o}[(Na,Pb,K,Li)_3(O,OH)_{30}]^{c}$	Orth. C222 ...	a=17.19Å Z=2 b=17.70Å c=3.933Å			Am.Min.,1970,55,1814(Abs.); Pov.,295-296;Hölzel,79.
RANKINITE	$Ca_3Si_2O_7$	$Ca_3{}^{[7]}\{g\}[Si_2{}^{t}O_7]$	Mon. $P2_1/a$	a=10.557Å β=119.59° b=8.885Å Z=4 c=7.858Å			SR,44A,314;SR,43A,320;Pov., 397;Hölzel,197.
RATHITE-I	$(Pb,Tl)_3As_5S_{10}$		Mon. $P2_1/a$	a=25,16Å β=100°28' b=7.94Å Z=4 c=8.47Å	$Pb_{I-III}(4e)$ $As_{I-V}(4e)$ $S_{I-X}(4e)$		Zeit.Krist.,1965,122,433-456; Pov.,235;Hölzel,42;K/S,74-75..
REINERITE	$Zn_3(AsO_3)_2$		Orth. Pbam	a=6.092Å Z=4 b=14.407Å c=7.811Å	$Zn_I(4f)$ $As_{II}(4g)$ $Zn_{II}(8i)$ $O_I(4h)$ $As_I(4h)$ ...		Am.Min.,1977,62,1129-1134; SR,43A,274-275;Pov.,289;Am. Min.,1959,44,207-208(Abs.).
RHODPLUMSITE	$Rh_3Pb_2S_2$		Trig. $R\bar{3}m$	a=5.73Å Z=3 c=14.00Å			Min.Abs.,84/892;Hölzel,16, 255.
ROBINSONITE	$Pb_4Sb_6S_{13}$	$Pb_4{}^{[6/8]}Sb_6{}^{[5/7]}[S_{13}]^{\#}$	Tric. P1	a=16.56Å α=91.09° b=17.69Å β=96.46° c=3.982Å γ=96.77° Z=2			Can.Min.,1982,20,97-100;SR, 45A,15;K/S,68-69;Hölzel,43.
ROMANÈCHITE	$(Ba,H_2O)_2Mn_5O_{10}$	$Mn_5{}^{o}[(Ba,H_2O)_2O_{10}]^{cchn}$	Mon. C2/m	a=13.929Å β=92.39° b=2.8459Å Z=2 c=9.678Å	$Mn_I(2b)$ $O_{I-VI}(4i)$ $Mn_{II-III}(4i)$ Ba(4i) (part.occ.)		Am.Min.,1988,73,1155-1161: Min.Abs.,87M/3126;Pov.,756; LF,108;Hölzel,74.
SAMARSKITE-(Y)	$(Y,Ce,U,Fe)_3(Nb,Ta,Ti)_5O_{16}$		Orth. C2/c	a=4.82Å Z=1 b=5.63Å (heated) c=5.15Å			Am.Min.,1985,70,856-866; Pov.,304;Str.Tab.,206;Hölzel, 77.
SARCOPSIDE	$(Fe,Mn,Mg)_3(PO_4)_2$	$(Fe,Mn,Mg)_3{}^{o}P_2{}^{t}[O_8]^{n}$ (relat. Forsterite)	Mon. $P2_1/a$	a=10.437Å β=90.0° b=4.768Å Z=2 c=6.026Å	$(Fe,Mn,Mg)_I(2a)$ $(Fe,Mn,Mg)_{II}(4e)$ P(4e) $O_{I-IV}(4e)$		Am.Min.,1972,57,24-35;SR, 38A,313;Pov.,537;K/B,119-120.
SCHNEIDERHÖH-NITE	$Fe_4As_5O_{13}$		Tric. $P\bar{1}$	a=8.924Å α=59.91° b=10.016Å β=112.41° c=9.103Å γ=81.69° Z=2	$As_{I-V}(2i)$ $Fe_{III-V}(2i)$ $Fe_I(1b)$ $O_{I-XIII}(2i)$ $Fe_{II}(1e)$		Can.Min.,1985,23,675-679: Am.Min.,1974,59,1139(Abs.): Hölzel,91.
SCHREYERITE	$V_2Ti_3O_9$		Mon.	a=7.06Å β=119.4° b=5.01Å Z=3 c=18.74Å			Am.Min.,1978,63,1182-1186; Hölzel,73.

Table 71

$A_pB_qC_r(p,q)\geq2;r\geq3$(cont.)

NAME	CHEMICAL FORMULA	STRUCTURAL FORMULA	SPACE GROUP	UNIT CELL DIMENSIONS	EQUIVALENT POSITIONS	STRUCTURE TYPE	REFERENCES
SEMSEYITE	$Pb_9Sb_8S_{21}$		Mon. C2/c	a=13.603Å β=106.0° b=11.935Å Z=4 c=24.452Å			Acta Cryst.,1974,B30,2935-2937;Min.Mag.,1969,37,442-446;Pov.,234;Str.Tab.,149.
SHANDITE	$Ni_3Pb_2S_2$		Trig. $R\bar{3}m$	a=5.576Å a_R=5.576Å c=13.658Å α=60° Z=3 Z_R=1	Pb_I(3a) S(6c) Pb_{II}(3b) Ni(9d)		Wyckoff,3,288-289;Am.Min.,1974,59,296-301;SR,45A,92-93;K/S,54-55;SR,13,278-379.
SINNERITE	$Cu_6As_4S_9$	$Cu_6{}^tAs_4{}^{[3n]}[S_9\square]^c$	Tric. P1	a=9.064Å α=90°00" b=9.830Å β=109°30' c=9.078Å γ=107°48' Z=2	As_{I-VIII}(1a) Cu_{I-XII}(1a) $S_{I-XVIII}$(1a)$\square_{I-II}$(1a)		Am.Min.,1975,60,998-1012; Am.Min.,1972,57,824-834.
SOPCHEITE	$Ag_4Pd_3Te_4$		Orth.	a=9.645Å Z=4 b=7.906Å c=11.040Å			Hölzel,suppl.;ZVMO,1982, 111,114-117.
SORBYITE	Pb_{19} $(Sb,As)_{20}S_{49}$		Mon. C2/m	a=45.15Å β=113°25' b=8.292Å (2b') Z=2 c=26.53Å			RRW,570;Pov.,253;Str.Tab.,149;Hölzel,45.
SPIROFFITE	$(Mn,Zn)_2$ Te_3O_8		Mon. Cc...	a=13.00Å β=98.22° b=5.38Å Z=4 c=12.12Å			Am.Min.,1962,47,196(Abs.); Pov.564-565;Str.Tab.,229; Hölzel,94.
SREBRO-DOLSKITE	$Ca_2Fe_2O_5$		Orth. Pnma	a=5.420Å Z=4 b=14.752Å c=5.594Å			Am.Min.,1986,71,1279-1280 (Abs,);Hölzel,59.
STANNO-MICROLITE	$(Sn,Fe,Mn)_2$ $(Ta,Nb,Sn)_2$ $(O,OH,F)_7$	$(Sn,Fe,Mn)_2{}^{cb}$ $(Ta,Nb,Sn)_2{}^{[6]}$ $[(O,OH,F)_7\square]^{Os}$	Cub. $Fd\bar{3}m$	a=10.57Å Z=8			Hölzel,67,274;Am.Min.,1972, 62,403-410;Min.Abs., 88M/1042,117.
STIBIO-MICROLITE	$(Ca,Sb,Na)_2$ $(Ta,Nb)_2$ $(O,OH,F)_7$	$(Ca,Sb,Na)_2{}^{cb}$ $(Ta,Nb)_2{}^{[6]}$ $[(O,OH,F)_7\square]^{Os}$	Cub. $Fd\bar{3}m$	a=10.455Å Z=8			Am.Min.,1987,73,1489(Abs.); Str.Tab.,192.
SUANITE	$Mg_2B_2O_5$		Mon. $P2_1/a$	a=12.10Å β=104°20' b=3.12Å Z=4 c=9.36Å			Pov.,472-473;Am.Min.,1963, 48,915-924;Hölzel,114;RRW, 590.
TALNAKHITE	$Cu_9(Fe,Ni)_8$ S_{16}	$Cu_9{}^t(Fe,Ni)_8{}^t$ $[S_{16},\square]^c$ (subs.deriv. Sphalerite)	Cub. $I\bar{4}3m$	a=10.593Å Z=2	(Fe,Ni)(2a) S_I(8c) (Fe,Ni)(6b) S_{II}(24g) Cu_I(12d) Cu_{II}(6g) ... (part.occ.)		Am.Min.,1972,57,368-380; Am.Min.,1970,55,2135(Abs.). Pov.229;Str.Tab.,115.
TENNANTITE	$(Cu,Fe)_{12}As_4$ S_{13}	$\{3\infty\}[(Cu,Fe)_6{}^t$ $(Cu,Fe)_6{}^{tr}As_4{}^{[3n]}$ $S_{13}]$	Cub. $I\bar{4}3m$	a=10.232Å Z=2	Cu_I(12d) S_{II}(2a) Cu_{II}(12e) As(8c) S_I(24g)	$\{3\infty\}[Cu,Fe)_6{}^t$ $(Cu,Fe))_6{}^{tr}Sb_4{}^{[3n]}S_{13}]$ TETRAHEDRITE	Zeit.Krist.,1966,123,1-20; K/S,4243;Hölzel,20;LF,283.
TETRADYMI-TE	$Bi_{14}Te_{13}S_8$	$Bi_{14}{}^o[Te_{13}S_8]^{chh}$ or $Bi_2{}^o[Te_2S]^{chh}$	Trig. $R\bar{3}$	a=4.25Å a_R=11.24Å c=29.60Å α=56.83° Z=3 Z_R=1	Bi_I(6c) Te_{II-II}(18f) Bi_{II-III}(18f) S_I(6c) Te_I(3a) S_{II}(18f)	$Bi_{14}{}^o[Te_{13}S_8]^{chh}$ or $Bi_2{}^o[Te_2S]^{chh}$ TETRADYMITE	Am.Min.,1975,60,994-997; Pov.,214-215;K/S,24-25; Hölzel,29;LF,102.
TETRAHE-DRITE	$(Cu,Fe)_{12}Sb_4$ S_{13}	$\{3\infty\}[(Cu,Fe)_6{}^t$ $(Cu,Fe)_6{}^{tr}Sb_4{}^{[3n]}S_{13}]$	Cub. $I\bar{4}3m$	a=10.364Å Z=2	$(Cu,Fe)_I$(12d) $(Cu,Fe)_{II}$(12e) Sb(8c) S_I(24g) S_{II}(2a)	$\{3\infty\}[(Cu,Fe)_6{}^t$ $(Cu,Fe)_6{}^{tr}Sb_4{}^{[3n]}S_{13}]$ TETRAHEDRITE	Min.Mag.,1986,50,717-721; Am.Min.,1988,73,389-397; Pov.,233;Str.Tab.,117;LF,283.

Table 72

$A_pB_qC_r(p,q)\geq 2;r\geq 3$(cont.)

NAME	CHEMICAL FORMULA	STRUCTURAL FORMULA	SPACE GROUP	UNIT CELL DIMENSIONS	EQUIVALENT POSITIONS	STRUCTURE TYPE	REFERENCES
THORTVEITI-TE	$(Sc,Y)_2Si_2O_7$	$(Sc,Y)_2^{o}\{g\}[Si_2^{t}O_7]$	Mon. C2/m	a=6.650Å β=102°20' b=8.616Å Z=2 c=4.686Å	(Sc,Y)(4h) Si(4i) O_{II}(4i) O_I(2a) O_{III}(8j)		Am.Min.,1988,73,601607;Str. Tab.,387;Pov.,396;SB,2,124; Acta Cryst.,1962,15,491-498.
TRITOMITE-(Y)	$(Y,Ca,Ce)_5(Si,B,Al)_3(O,OH)_{13}$		Hex.	a=9.35Å c=6.88Å Z=2			Am.Min.,1962,47,925;RRW, 628;Hölzel,194;Pov.,762,389-390;LF,165.
TUČEKITE	$Ni_9Sb_2S_8$		Tet. P4/mmm	a=7.174Å Z=1 c=5.402Å			Min.Mag.,1978,42,278;Am.Min., 1979,64,465(Abs.);Hölzel,17.
TVALCHRELI-DZEITE	$Hg_{12}(As,Sb)_8S_{12}$		Tric. P1	a=4.391Å α=88.17° b=11.573Å β=90.01° c=15.667Å γ=89.98° Z=1			Dokl.Earth Sci.,1986,290, 1208-1212;Hölzel,39;Am.Min., 1977,62,174(Abs.).
TVEITITE-(Y)	$Ca_{14}Y_5F_{43}$		Trig. $R\bar{3}$ ...	a=16.699Å a_R=? c=9.666Å ... Z=3	Y(18f) Ca_I(6c) Ca_{II-III}(18f) F_I(6c) $F_{II-VIII}$(18f) (part.occ.) (for $Ca_{13}Y_6F_{43}$)		J.Sol.St.Chem.,1982,44,75-81;Am.Min.,1977,62,1060 (Abs.);Hölzel,51.
VEENITE	$Pb_2(Sb,As)_2S_5$		Orth. $P2_1cn$ ...	a=8.44Å (2a') Z=8 b=26.2Å c=7.90Å			Can.Min.,1967,9,7-24; RRW,651;Pov.,235;Hölzel,42.
XANTHIOSITE	$Ni_3(AsO_4)_2$		Mon. $P2_1/a$	a=10.174Å β=92°58' b=9.548Å Z=4 c=5.766Å			Min.Mag.,1963,35,72-83; Pov.,509:Hölzel,143.
XILINGOLITE	$Pb_{3.3}Bi_{1.8}S_6$		Mon. C2/m ...	a=13.65Å β=93.0° b=4.078Å Z=4 c=20.68Å			Am.Min.,1984,69,409(Abs.); N.Jb.Min.Abh.,1989,160,260-297;Hölzel,44.
YTTRIALITE-(Y)	$(Y,Th)_2Si_2O_7$ α-form	$(Y,Th)_2^{o}\{g\}[Si_2^{t}O_7]$	Mon. $P2_1/m$	a=7.50Å β=112°0' b=8.06Å Z=2 c=5.02Å (heated)	$(Y,Th)_I$(2a) $(Y,Th)_{II}$(2b) Si_{I-II}(2e) ...		Sov.Phys.Cryst.,1972,16,786-789;SR,37A,346;Pov.,396;Min. Abs.,75-303,29;Hölzel,197.
ZIESITE	$Cu_2V_2O_7$ β-form	$Cu_2^{[5]}\{g\}[V_2^{t}O_7]$	Mon. C2/c	a=7.6890Å β=110.252° b=8.0289Å Z=4 c=10.1065Å	V(8f) O_I(4e) Cu(8f) O_{II-IV}(8f)		N.Jb.Min.Mh.,1989,41-47; Am.Min.,1980,65,1146-1149; Hölzel,143.
ZINKENITE	$Pb_9Sb_{22}S_{42}$		Hex. $P6_3$	a=22.108Å Z=1 c=4.3265Å	Pb_I(6c)Pb_{II-III}(2a) (Pb,Sb)(6c) Sb_{I-III}(6c) S_{I-VII}(6c) (part.occ.)		Am.Min.,1986,71,194-201; Min.Abs.,89M/0908;Zeit.Krist., 1975,141,79-96;K/S,69-70.

2.2. The layer description of close-packed structures. The use of condensed models and the systematic derivation of mineral structures

For the close-packed minerals the layer description is presented, using condensed models (Tables 1L to 17L). When the layers are all alike only one layer is represented. If the structure is built of different layers only the different ones are shown, but the sequence of the layers is always expressed (e.g., tellurobismuthite, $Bi_2^{O}[Te_3]^{chh}$).

One can realize that the patterns of either the packing or of the interstitial atoms are the most symmetrical for the corresponding proportion of occupation (see also Lima-de-Faria, 1994). The layer description is very simple, and obeys simple stacking rules (Lima-de-Faria, 1978).

The systematic derivation of the crystal structures on the basis of physico-chemical principles has been one of the author's interests from very early on (Lima-de-Faria, 1965, and Lima-de-Faria and Figueiredo, 1969). Based on the simplicity of the results of the layer description one can try to derive the various close-packed mineral structures. However, such a derivation will not be undertaken here.

Table 1L

	A
OSMIUM $[Os]^{h}$ $P6_3/mmc$ **COPPER** $[Cu]^{c}$ $Fm\bar{3}m$	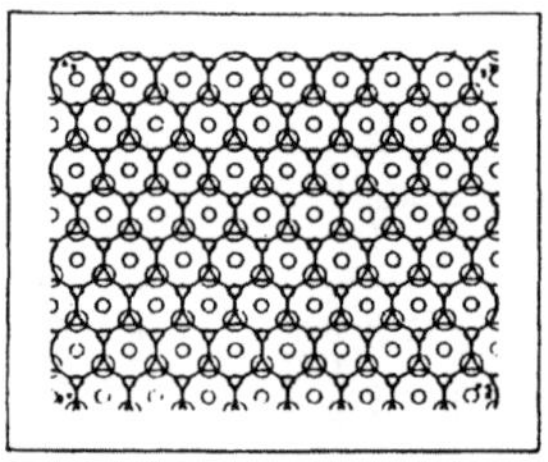
IRON $[Fe]^{b}$ $Im\bar{3}m$	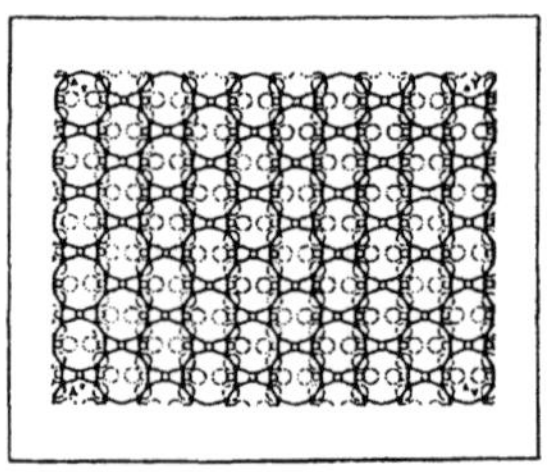

Table 2L

AB

[XY]

TETRAAURICUPRIDE (Cuproaurite) $[CuAu]^{c}$ P4/mmm (subs. deriv. Copper)	pack.  1/2	
HONGSHIITE $[PtCu]^{c}$ R32... (subs. deriv. Copper)	pack. 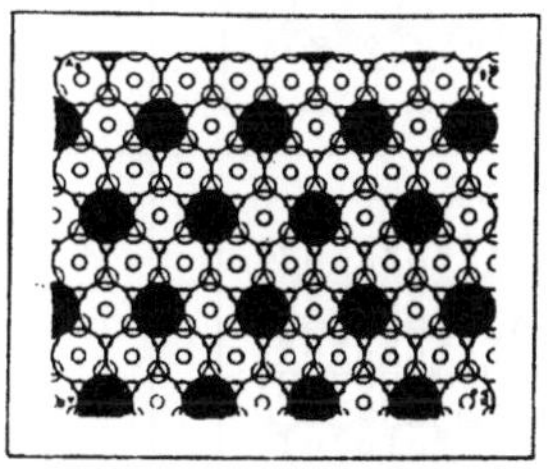 1/3	pack.  2/3
WAIRAUITE $[CoFe]^{b}$ $Pm\bar{3}m$ (subs. deriv. Iron)	pack. 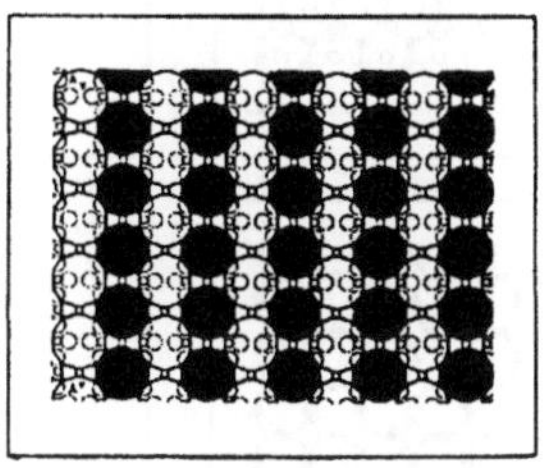 1/2	
CAROBBIITE $[KF]^{Qs}$ $Fm\bar{3}m$	pack. 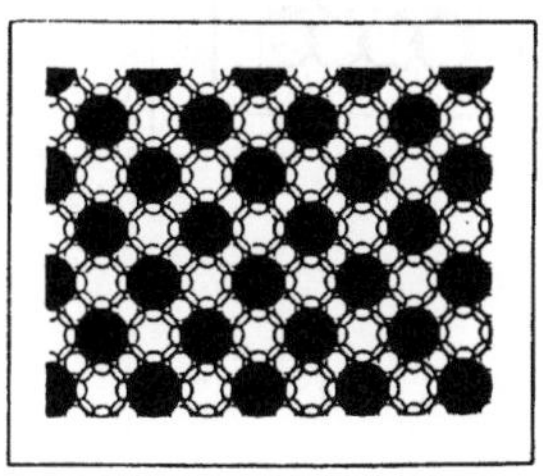 1/2	

Table 3L

AB(cont.)

A[X]

NICKELINE (Niccolite) $Ni^{o}[As]^{h}$ $P6_3/mmc$ **HALITE** $Na^{o}[Cl]^{c}$ $Fm\overline{3}m$	oct. 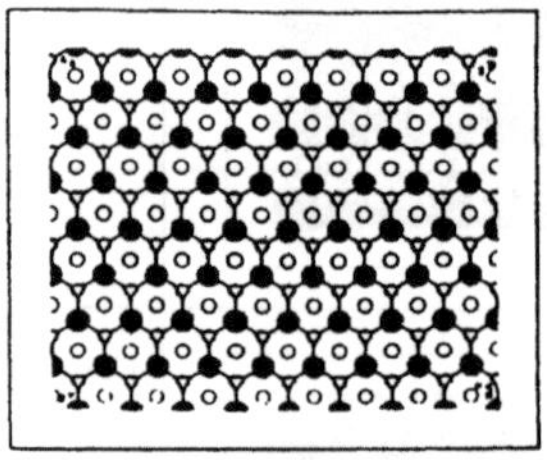 1	
WURTZITE-2H $Zn^{t}[S]^{h}$ $P6_3mc$ **SPHALERITE** $Zn^{t}[S]^{c}$ $F\overline{4}3m$	t 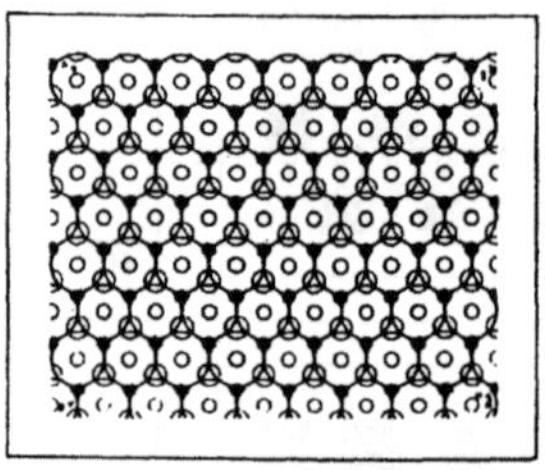 1/2	
LITHARGE $O^{t}[Pb]^{c}$ $P4/nmm$	t 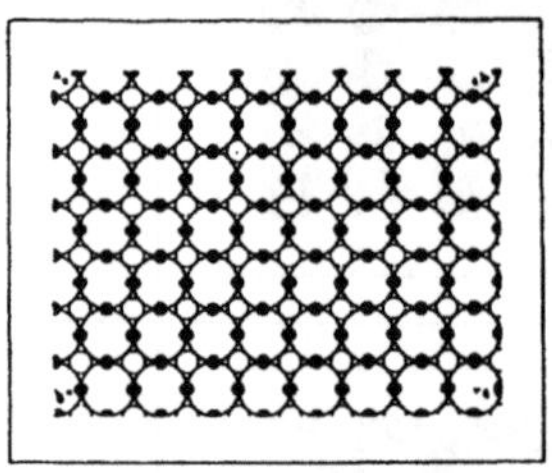 1	t 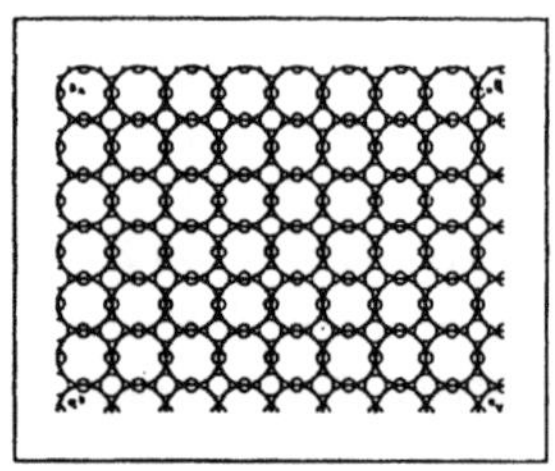 0

Table 4L

AB(cont.)

A[X](cont.)

SAL AMMONIAC $(NH_4)^{cb}[Cl]^{Qs}$ $Pm\bar{3}m$	cb 1	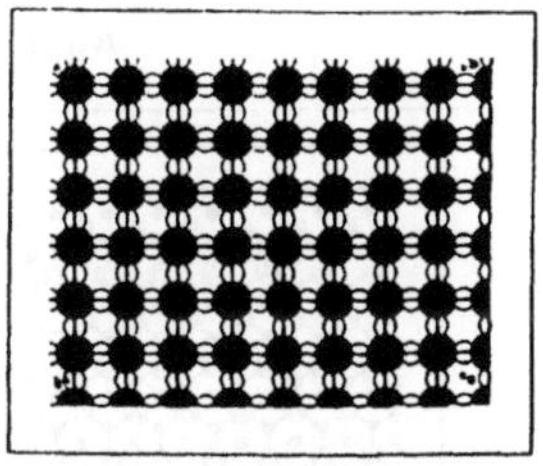
COOPERITE $(Pt,Pd,Ni)^{sq}[S]^{Qs}$ $P4_2/mmc$	sq 1/3	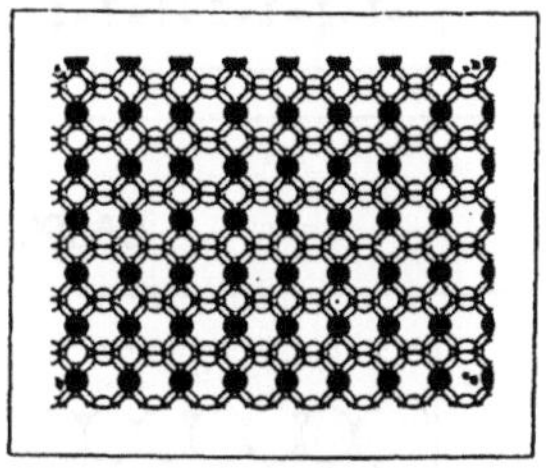
VYSOTSKITE $(Pd,Ni)^{sq}[S]^{N21\,s}$ $P4_2/m$	sq 1/3	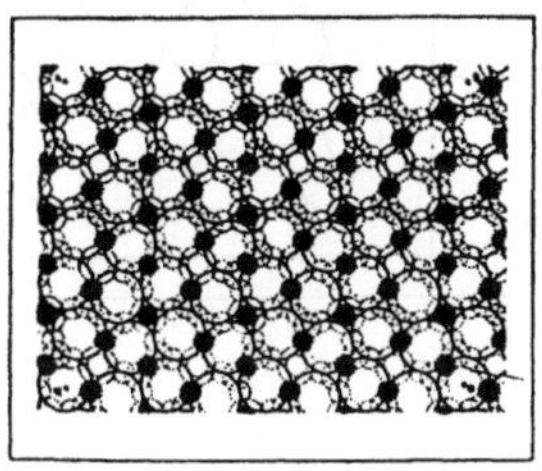

Table 5L

AB_2

$A_2[X]$

CHALCOCITE $Cu_2^{tr/t}[S]^h$ $P2_1/c$ **DIGENITE** $Cu_2^{t/tr}[S]^c$ or $(Cu_9,\square)^{t/tr}[S_5]^c$ $R\bar{3}m$	t 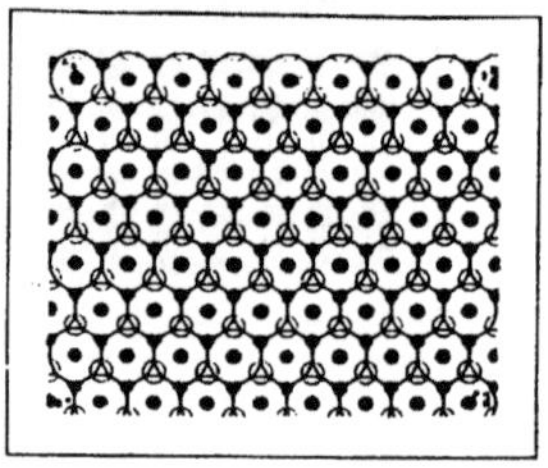 1

$[[A_2][X]]$

MALDONITE $[[Au_2][Bi]]^{(.-TK)\ c}$ $Fd\bar{3}m$	K 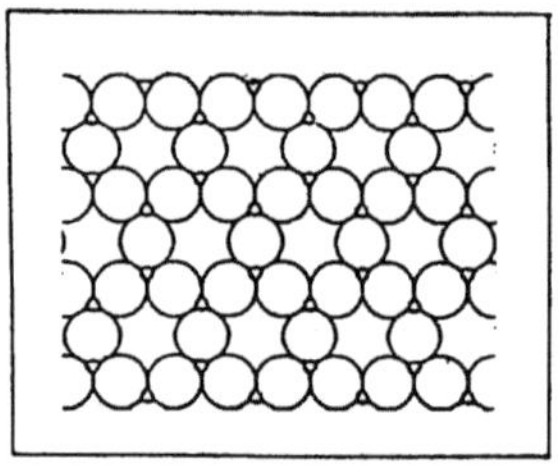	$\dot{\tilde{T}}$ 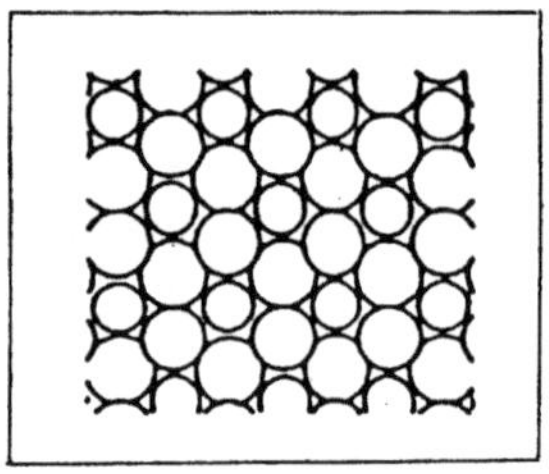

Table 6L

AB_2(cont.)

$A[X_2]$

RUTILE $Ti^{o}[O_2]^{h}$ $P4_2/mnm$	oct. 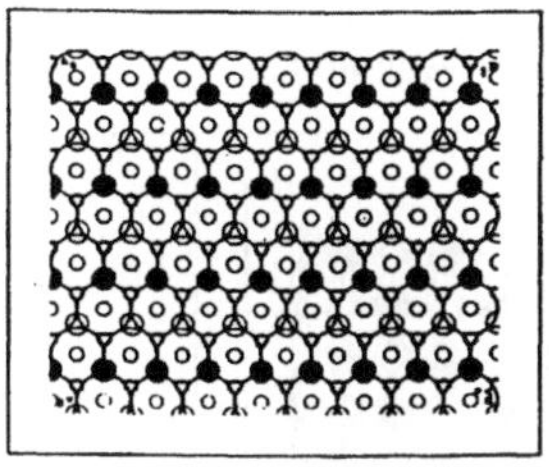 1/2	
SCRUTINYITE $Pb^{o}[O_2]^{h}$ Pcan **ANATASE** $Ti^{o}[O_2]^{c}$ $I\ 4_1/amd$ **BROOKITE** $Ti^{o}[O_2]^{ch}$ Pbca	oct. 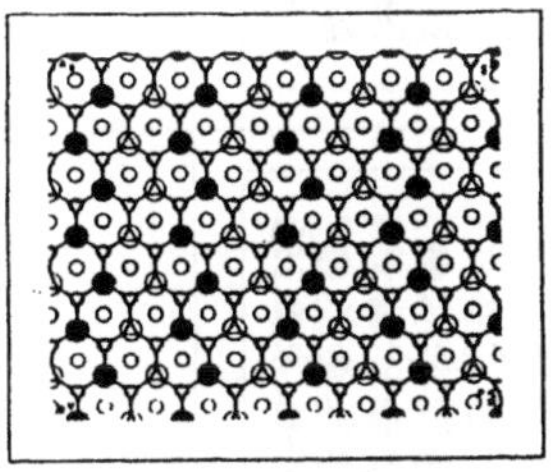 1/2	
RAMSDELLITE $Mn^{o}[O_2]^{h}$ Pbnm	oct. 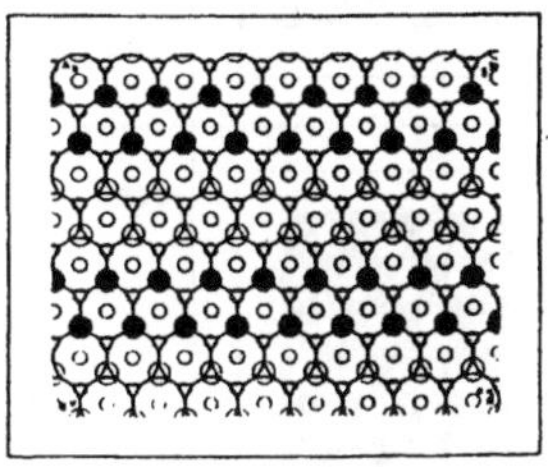 1/2	
MELONITE $Ni^{o}[Te_2]^{h}$ $P\bar{3}m1$ **CHLOROMAGNESITE** $Mg^{o}[Cl_2]^{c}$ $R\bar{3}m$	oct. 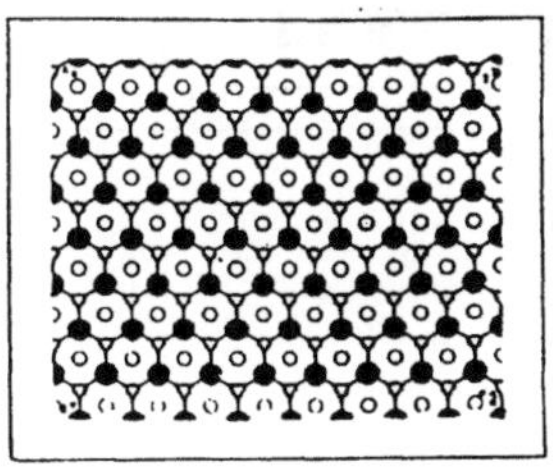 1	oct. 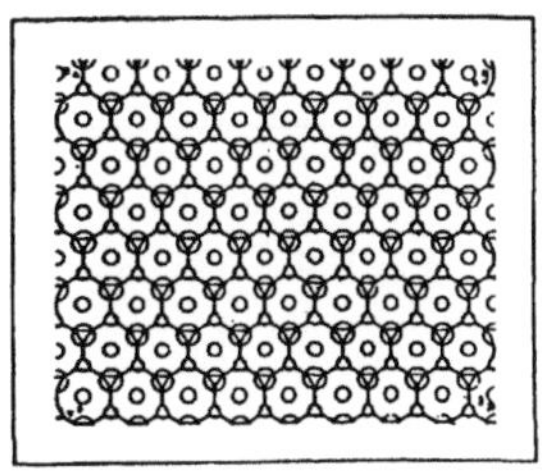 0

Table 7L

AB_2(cont.)

$A[X_2]$(cont.)

FLUORITE $Ca^{cb}[F_2]^{Qs}$ $Fm\bar{3}m$	cb 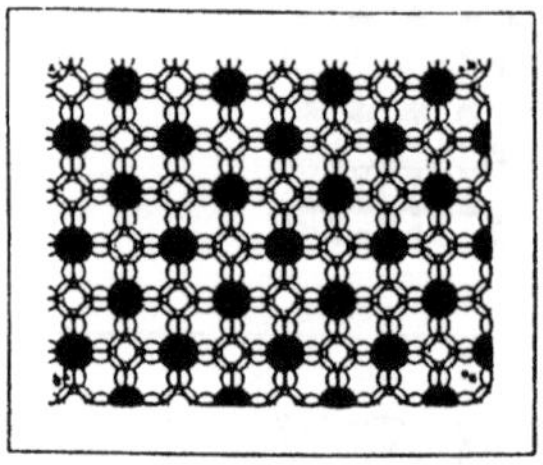 1/2	
ANYUIITE $Au^{acb}[Pb_2]^{N21\,f}$ I 4mcm	acb 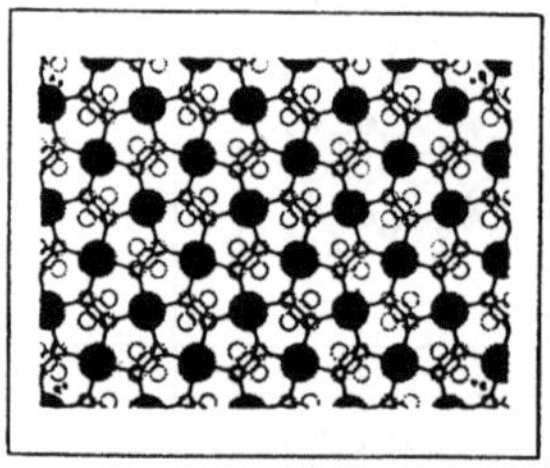 1	
MOLYBDENITE-2H $Mo^{P}[S_2]^{(2Ts)h}$ $P6_3/mmc$ **MOLYBDENITE-3R** $Mo^{P}[S_2]^{(2Ts)c}$ $R\bar{3}m$	oct. 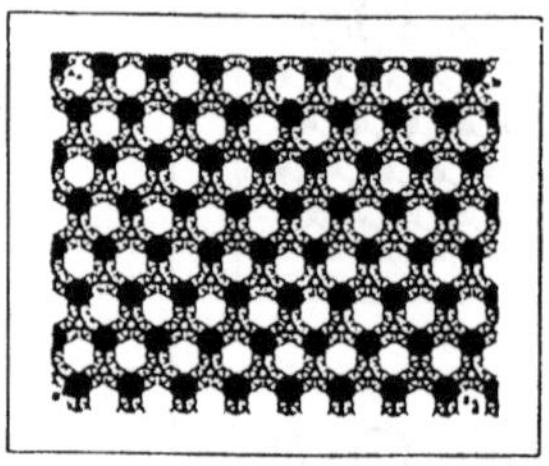 1	oct. 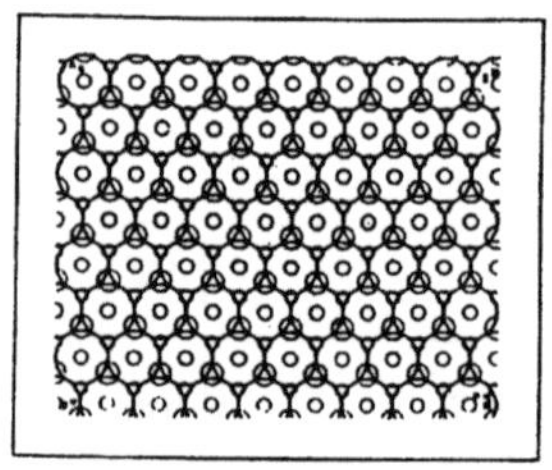 0

Table 8L

AB_3

$[XY_3]$

AURICUPRIDE $[Cu_3Au]^c$ $Pm\bar{3}m$ (subs. deriv. Copper)	pack. 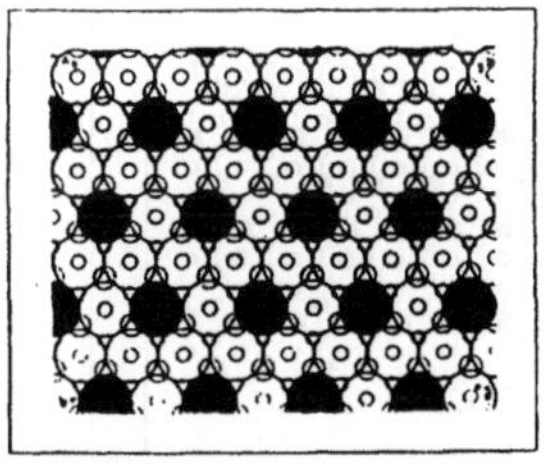 1/4	
DYSCRASITE $[Ag_3Sb]^h$ Pmm2 (subs. deriv. Osmium)	pack.  1/2	pack. 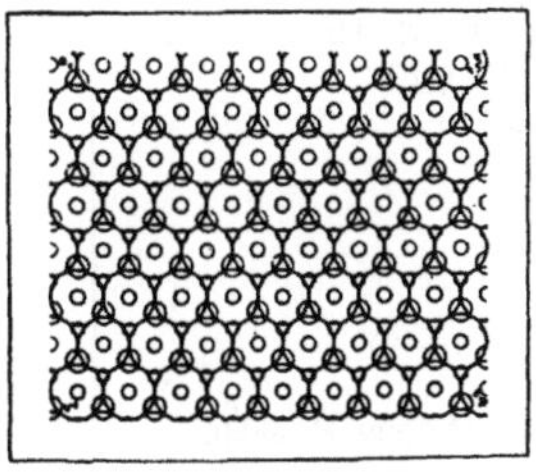 0

Table 9L

AB_3(cont.)

A[X_3]

MOLYSITE $Fe^{o}[Cl_3]^{h}$ R$\bar{3}$ **MOLYBDITE** $Mo^{o}[O_3]^{c}$ Pbnm	oct. 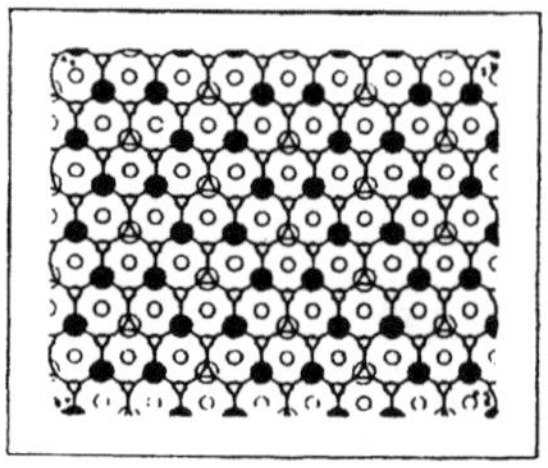 2/3	oct. 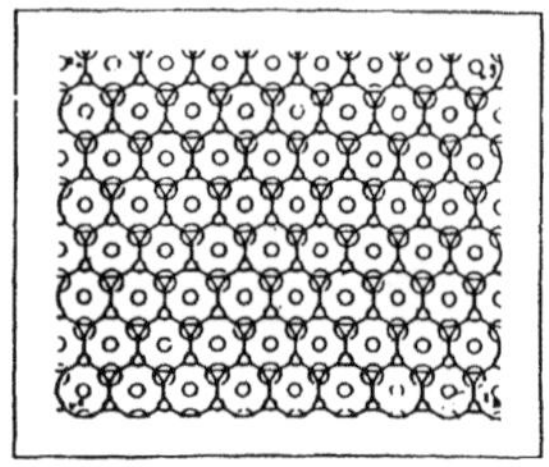 0
GIBBSITE $Al^{o}[(OH)_3]^{(2Ts)h}$ P2_1/n	oct. 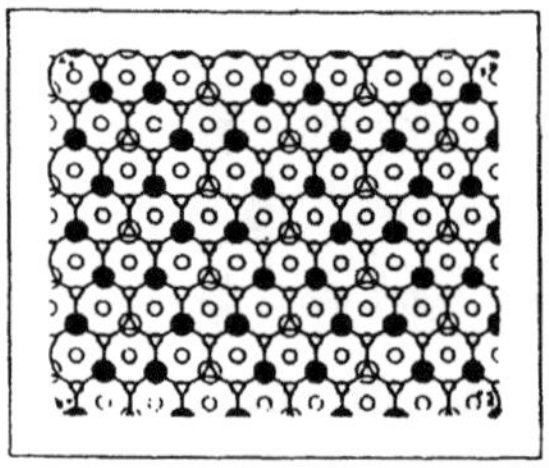 2/3	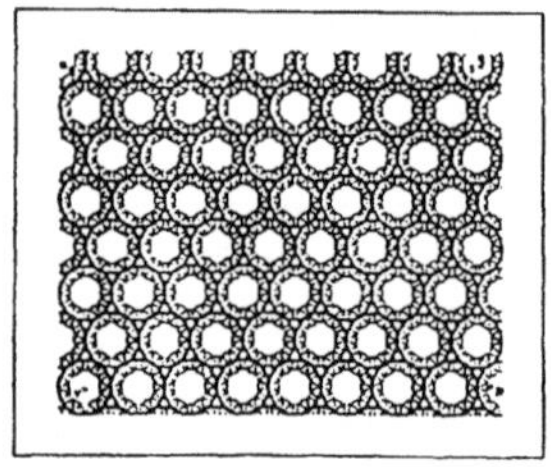 0
DZHALINDITE $In^{o}[(OH)_3\square]^{c}$ Im$\bar{3}$ (defect deriv. Roaldite)	oct. □ 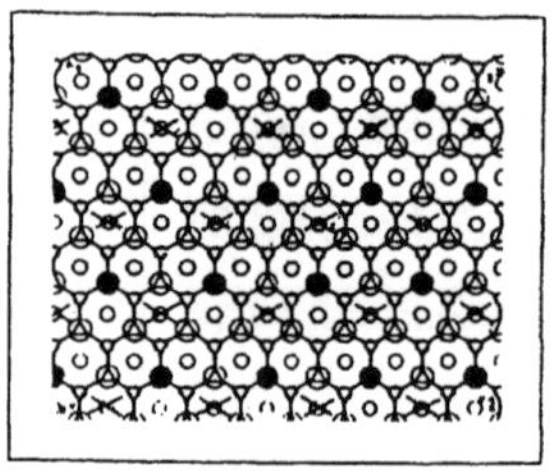 1/4 !/4	

AB_4

A[X_4]

ROALDITE $N^{o}[(Fe,Ni)_4]^{c}$ Pm$\bar{3}$m	oct. 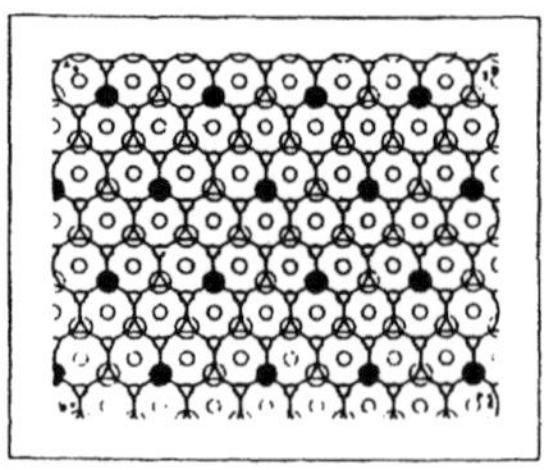 1/4

Table 10L

A_2B_3

CORUNDUM $Al_2^{o}[O_3]^{h}$ $R\bar{3}c$	oct. 2/3 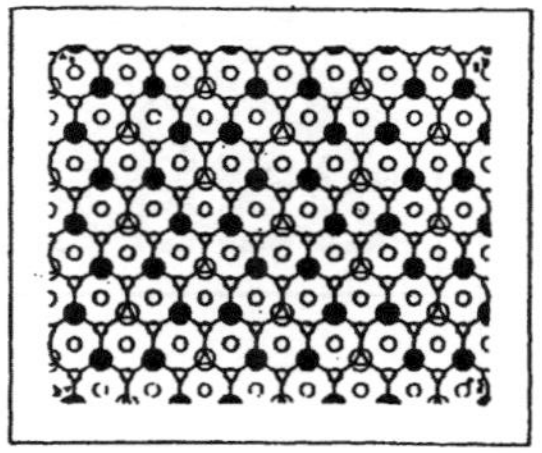		
TELLUROBISMU-THITE $Bi_2^{o}[Te_3]^{chh}$ $R\bar{3}m$	oct. 1 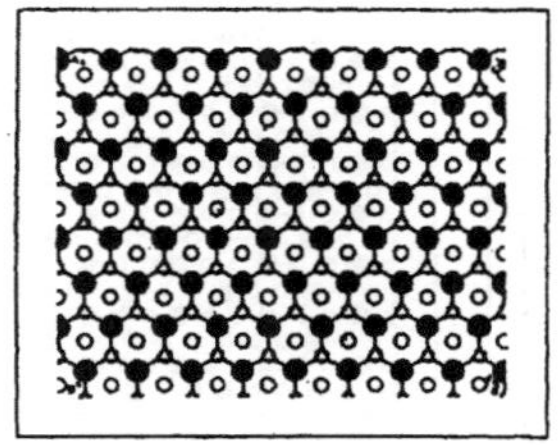	oct. 1 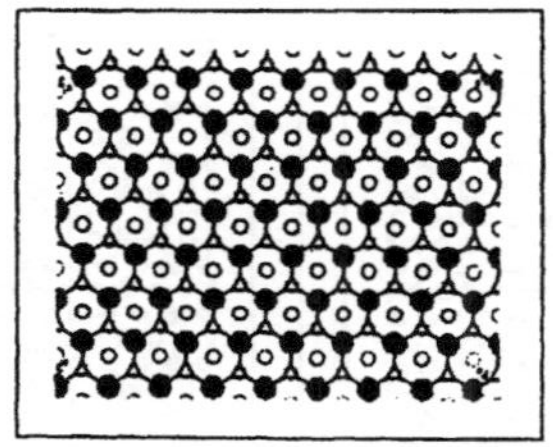	oct. 0 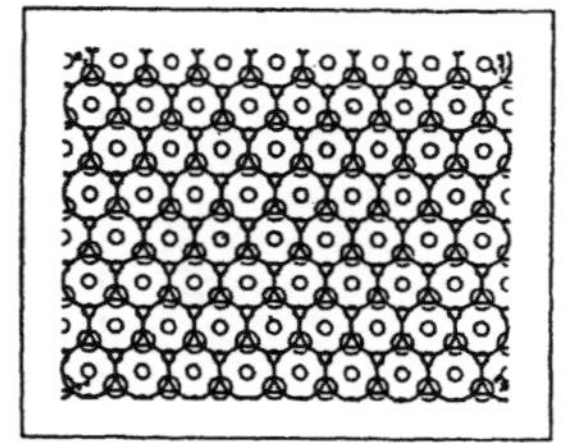
BIXBYITE $Mn_2^{[6]}[O_3\square]^{Qs}$ $Ia\bar{3}$ (defect deriv. Fluorite)	cb □ 1/2 1/4 		

Table 11L

ABC

MANGANITE $Mn^{o}[O(OH)]^{h}$ $B2_1/d$ (subs.deriv. Rutile)	oct. 1/2	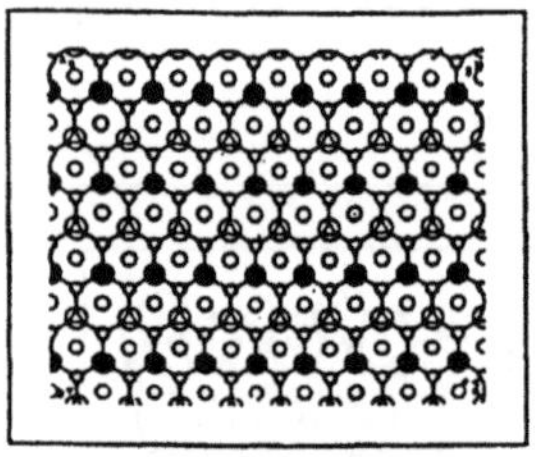		
DIASPORE $Al^{o}[O(OH)]^{h}$ Pbnm (subs.deriv. Ramsdellite)	oct. 1/2	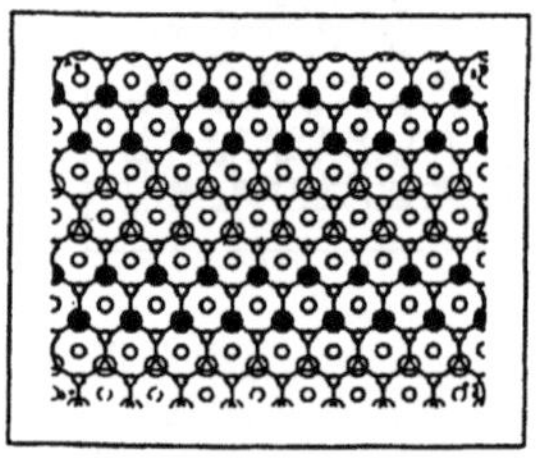		
HETEROGENITE-2H $Co^{o}[O(OH)]^{(2Ts)h}$ $P6_3/mmc$ **HETEROGENITE-3R** $Co^{o}[O(OH)]^{(2Ts)c}$ $R\bar{3}m$	oct. 1	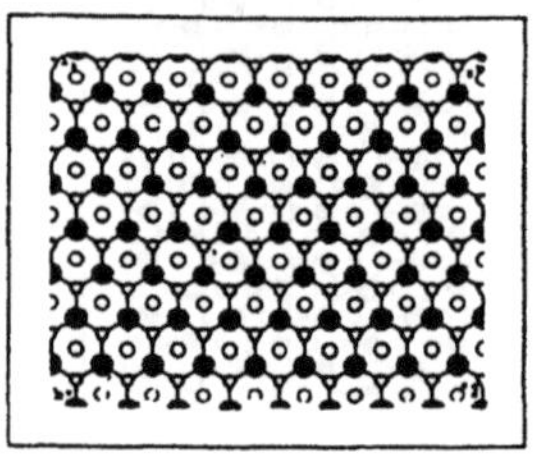	0	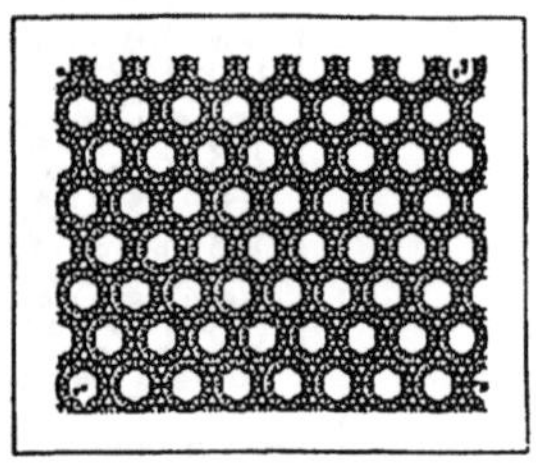

O and OH are not discriminated.

Table 12L

ABC(cont.)

MATLOCKITE $Pb^{[9]}[[F][Cl]]^{(Q2,2Q1\ f)\ f}$ P4/nmm	[9] 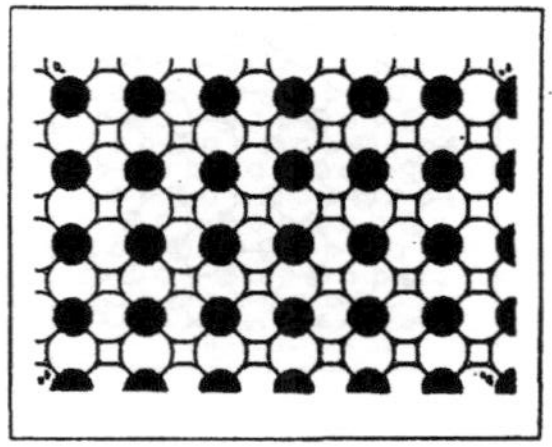 1	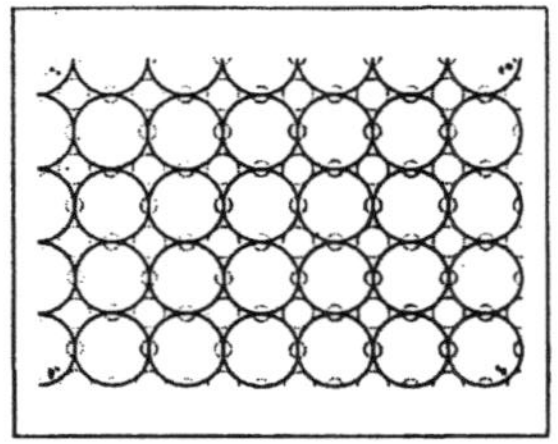 0	[9] 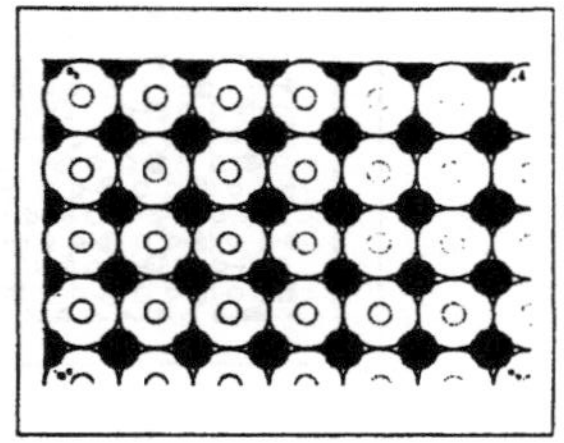 1

ABC_2

MATILDITE $Ag^{o}Bi^{o}[S_2]^{c}$ $P\bar{3}m1$ (subs. deriv. Halite)	oct. 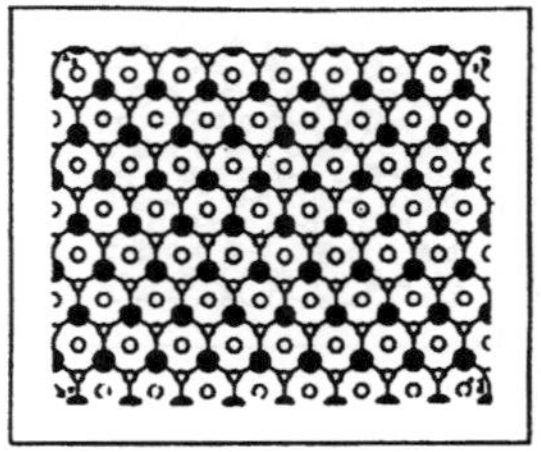 1	oct. 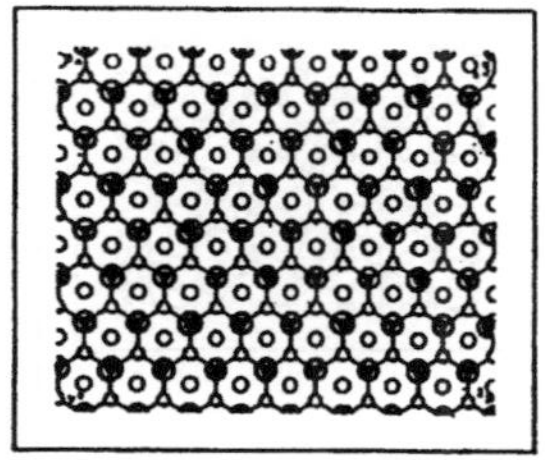 1	
CHALCOPYRITE $Cu^{t}Fe^{t}[S_2]^{c}$ $I\bar{4}2d$ (subs. deriv. Sphalerite)	t 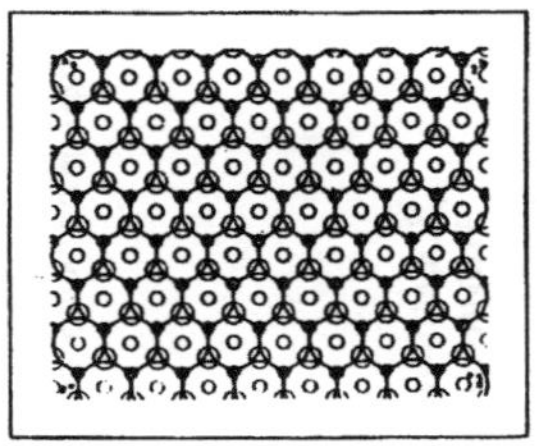 1/2		

Table 13L

ABC_3

ILMENITE $Fe^{o}Ti^{o}[O_3]^{h}$ $R\bar{3}$ (subs. deriv. Corundum)	oct. 2/3	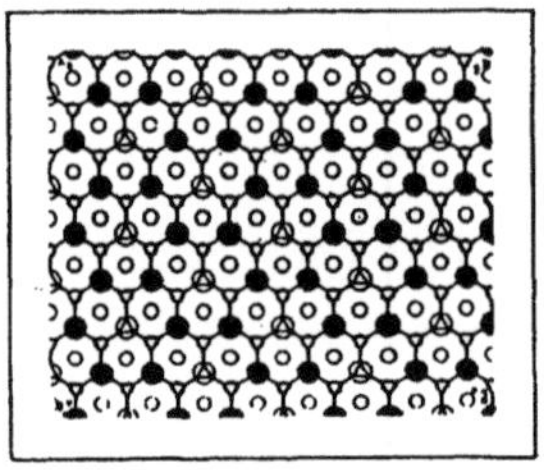	oct. 2/3	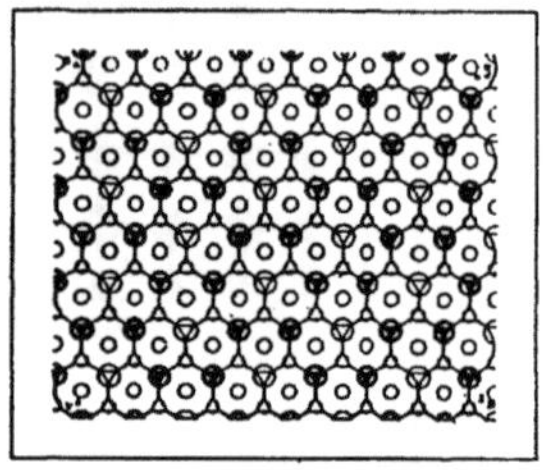
PEROVSKITE $Ti^{o}[CaO_3]^{c}$ Pbnm (subs. deriv. Roaldite)	oct. 1/4	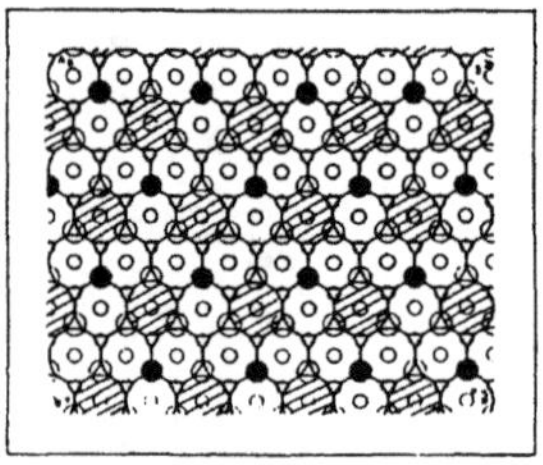		

ABC_4

WOLFRAMITE $(Fe,Mn)^{o}W^{o}[O_4]^{h}$ P2/c (subs. deriv. Scrutinyite)	oct. 1/2	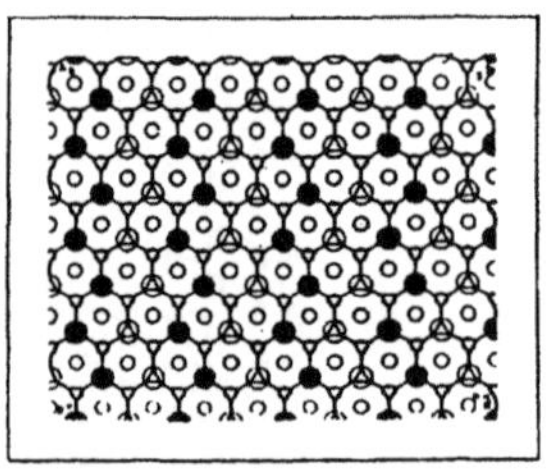	oct. 1/2	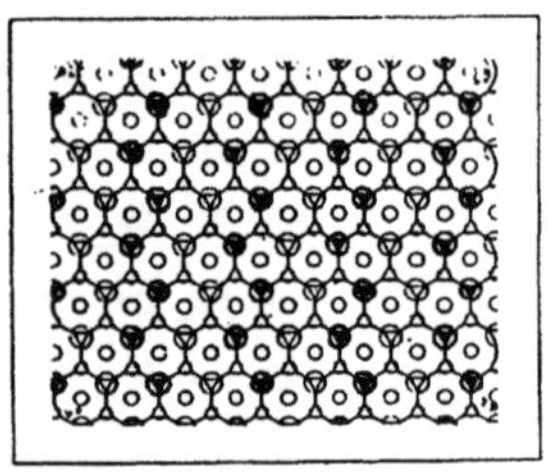
HETEROSITE $(Fe,Mn)^{o}P^{t}[O_4]^{h}$ Pmnb (relat. Forsterite)	oct., t 1/4 , 1/8	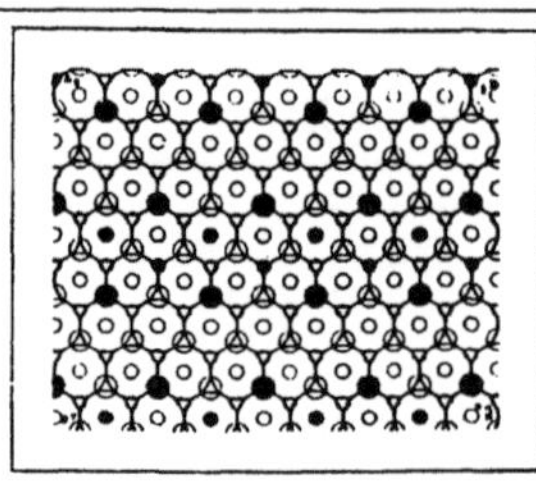		

Table 14L

AB_2C_2

TETRADYMITE $Bi_2{}^{o}[Te_2S]^{chh}$ or $Bi_{14}{}^{o}[Te_{13}S_8]^{chh}$ $R\bar{3}$ (subs. deriv. Tellurobismuthite)	oct. 1 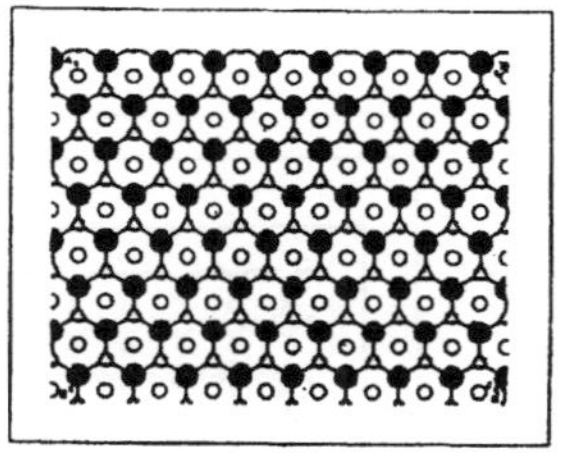	oct. 1 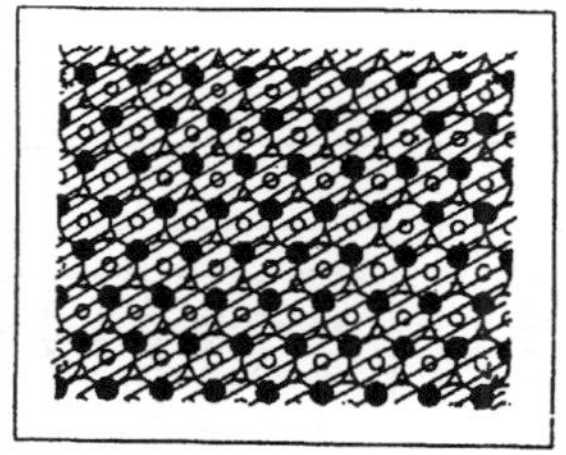	oct. 0 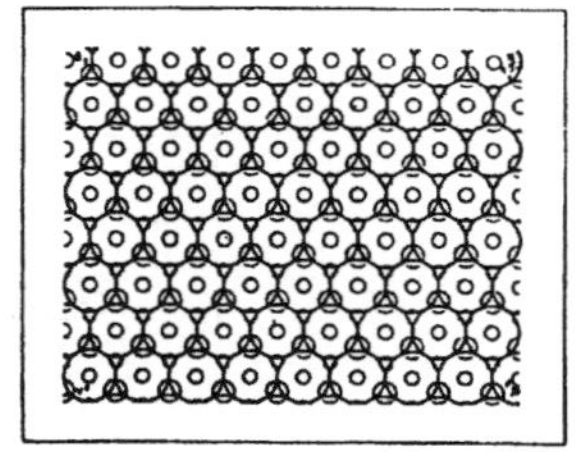

AB_2C_3

CUBANITE $Cu^{t}Fe_2{}^{t}[S_3]^{h}$ Pcmn	t 1/2 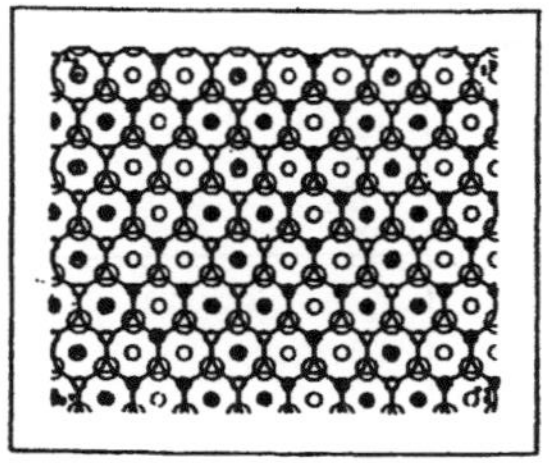		
COVELLITE (Covelline) $Cu_2{}^{t}Cu^{tr}[S_3]^{(3Th)\,s}$ $P6_3/mmc$	t 1/2 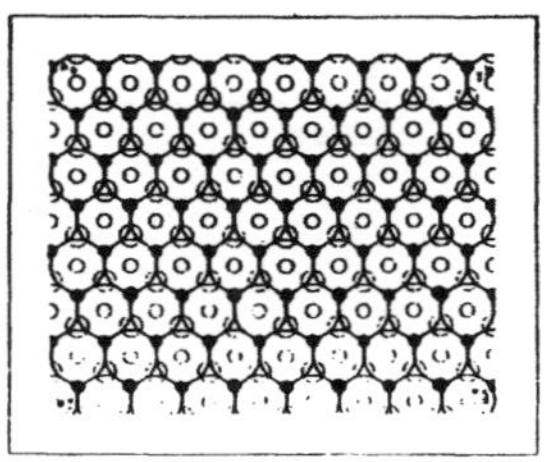	t , tr 1/2 , 1/2 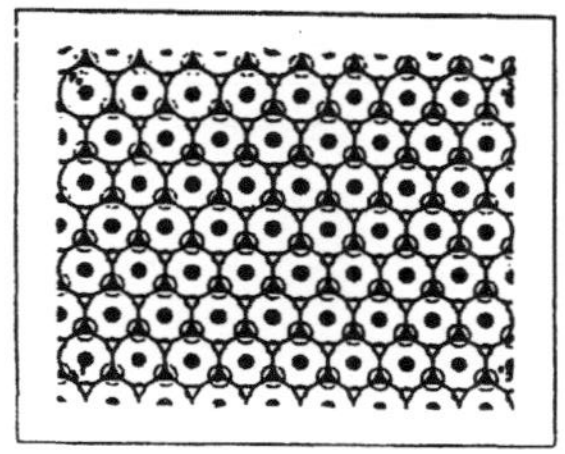	0 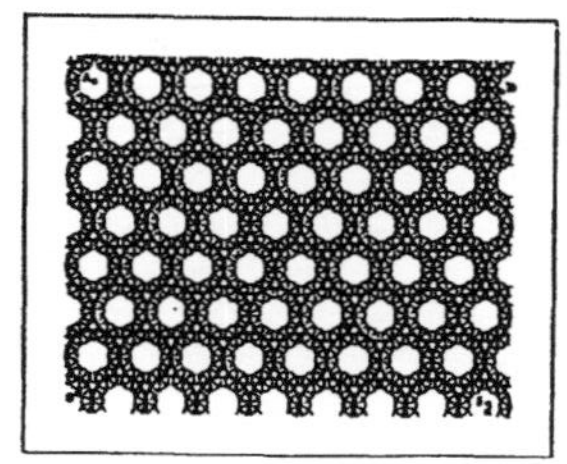

Table 15L

AB_2C_4

FORSTERITE $Mg_2{}^{o}Si^{t}[O_4]^{h}$ Pnma	oct., t 1/2 , 1/8	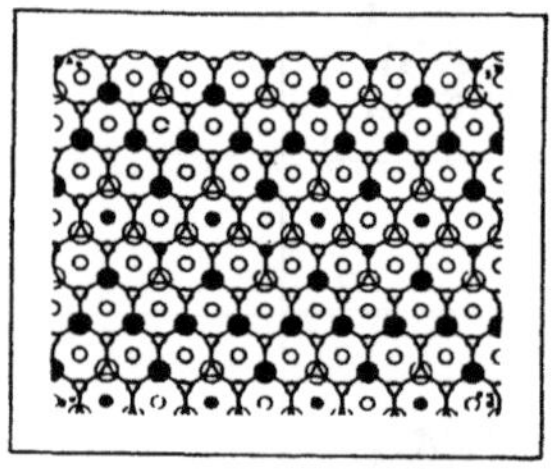
SPINEL $Al_2{}^{o}Mg^{t}[O_4]^{c}$ $Fd\bar{3}m$	oct., t 1/2 , 1/8	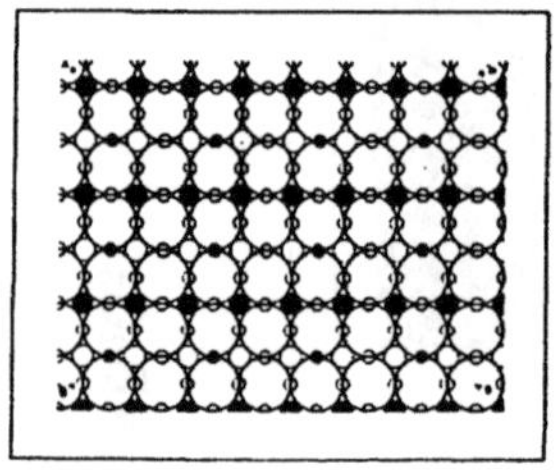

AB_2C_5

KYANITE $Al_2{}^{o}Si^{t}[O_5]^{c}$ $P\bar{1}$	oct., t 2/5 , 1/10	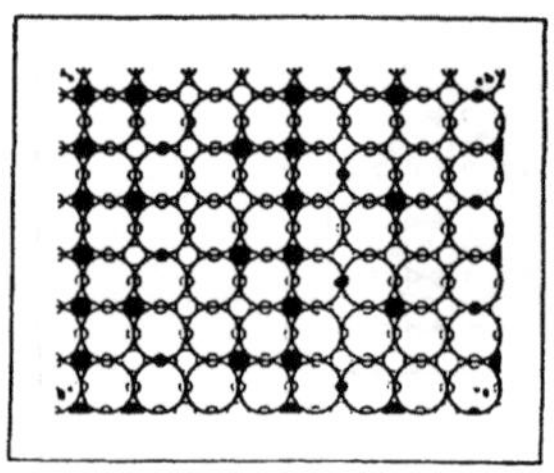

Table 16L

AB_2C_6

TAPIOLITE $(Fe,Mn)^{o}(Ta,Nb)_2^{o}[O_6]^{h}$ $P4_2/mnm$ (subs. deriv. Rutile)	oct. 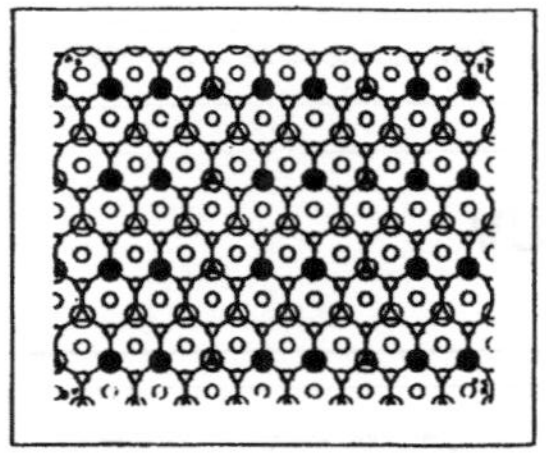 1/2		
TANTALITE $(Fe,Mn)^{o}Ta_2^{o}[O_6]^{h}$ Pbcn. (subs. deriv. Scrutinyite)	oct. 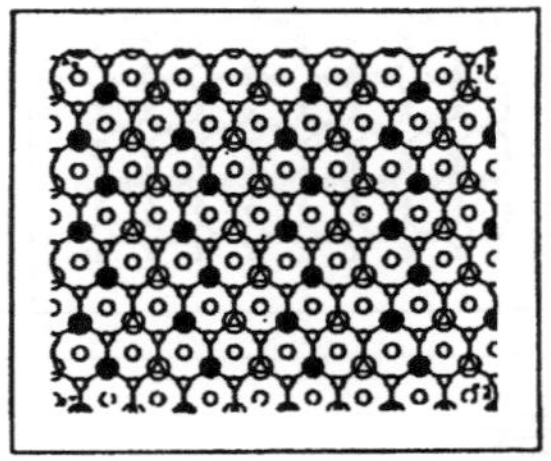 1/2	oct. 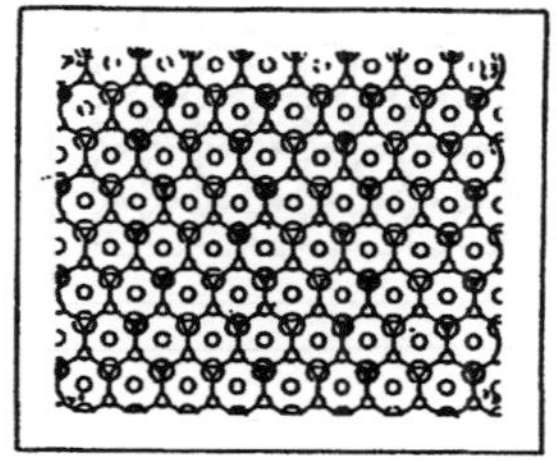 1/2	oct. 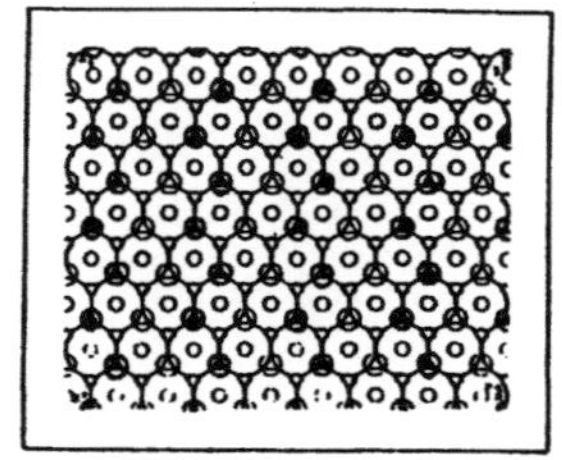 1/2
HIERATITE $Si^{o}[K_2F_6]^{c}$ $Fm\bar{3}m$ (relat. Perovskite)	oct. 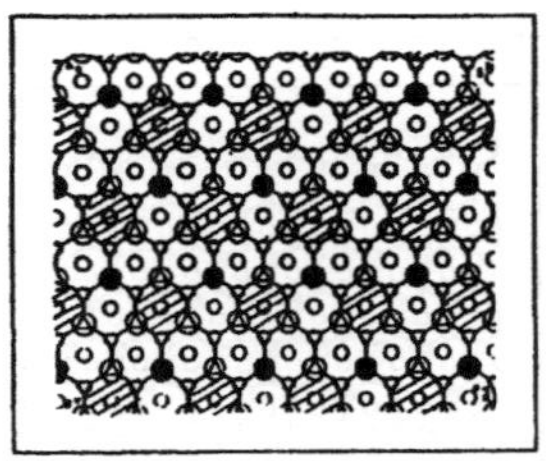 1/4	oct. 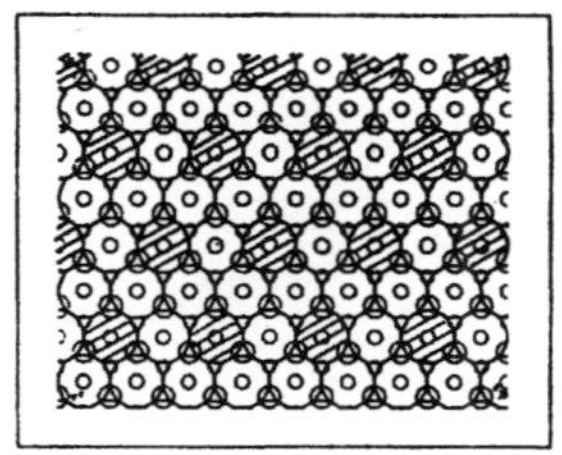 0	

Table 17L

AB_3C_4

ENARGITE $Cu_3^{t}As^{t}[S_4]^{h}$ $Pnm2_1$ (subs. deriv. Wurtzite) **FAMATINITE** $Cu_3^{t}Sb^{t}[S_4]^{c}$ $I\,\bar{4}2m$ (subs. deriv, Sphalerite)	t 1/2	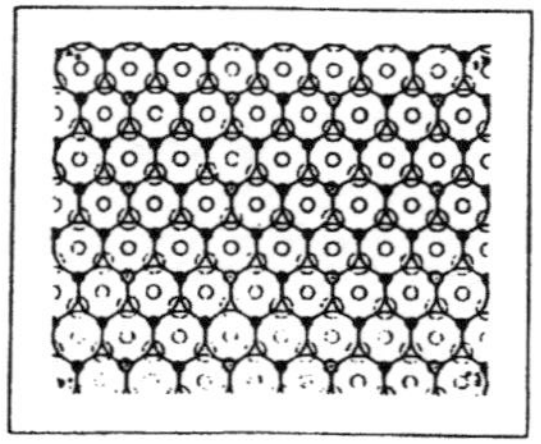

AB_8C_8

PENTLANDITE $(Ni,Fe)^{o}(Ni,Fe)_8^{t}[S_8]^{c}$ $Fm\bar{3}m$	oct., t 1/4, 1/2	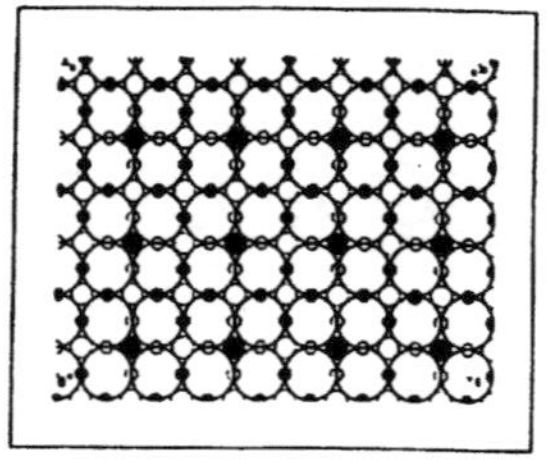	oct., t 0 , 1/2	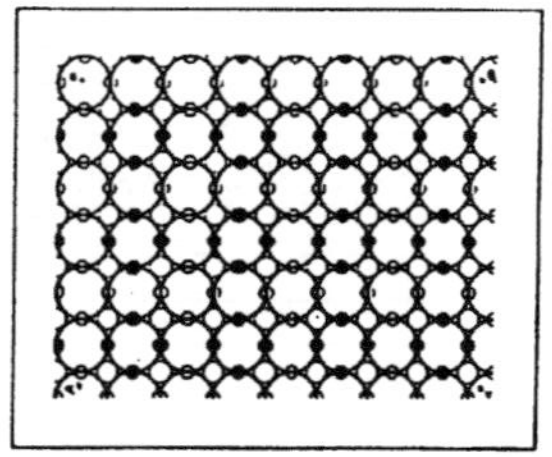

AB_8C_{16}

HOLLANDITE $Mn_8^{o}[(Ba,K)\,O_{16}]^{chh}$ $I\,2/m$	oct. 1/3	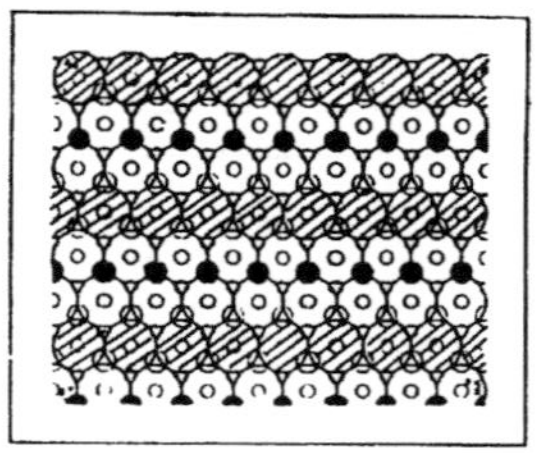	oct. 2/3	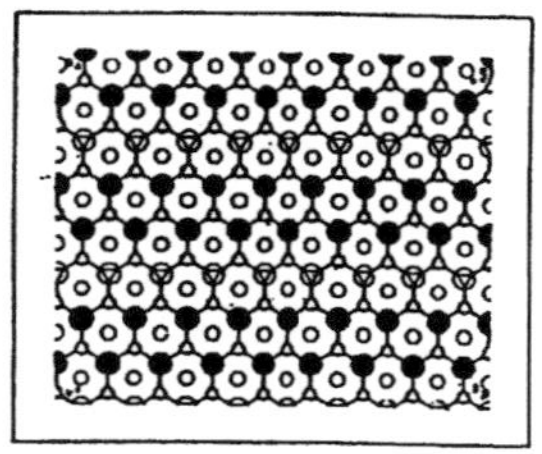	oct. 1/3	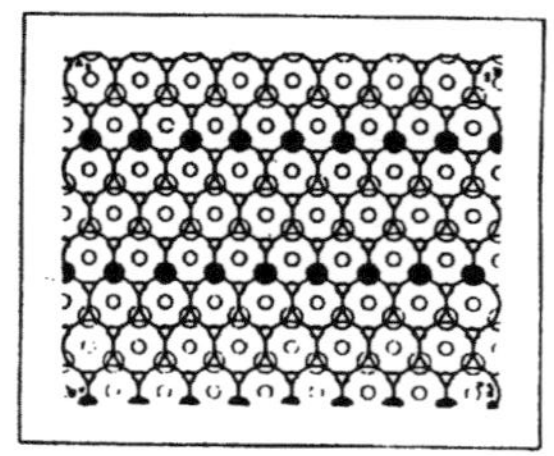

2.3. The tables of mineral structure types

The mineral species are here organized in structure types, in order to show how many different atomic arrangements exist (Tables 1S to 22S), and some summarizing results are presented in Tables 23S to 26S.

Like chemical elements the mineral structure types call for some display work, such as a map or a table (Table 23S). The aim is to have a global view of the main mineral structure types in order to understand how the chemical elements are linked together to form minerals, and to reveal the rules, laws, tendencies, principles etc., of the architecture of the mineral kingdom. The main close-packed mineral structure types of this study are summarized in Table 26S. For a global approach to this problem see the general table of inorganic structure types by Lima-de-Faria and Figueiredo (1976, 1978) (see also Lima-de-Faria, 1994). Table 26S shows the relationships between the main kinds of close packings and their substitution and insertion derivatives.

About the chemical elements Spronsen (1969) wrote: "There are three distinguishable main stages in the history of the periodic system, namely, that of initiation, that of phenomenological development and that of theoretical development". The table of inorganic structure types, and in particular that of mineral structure types, has possibly to go through similar stages. We are now in the first two stages, which will perhaps stimulate the beginning of the theoretical development, by trying to explain why certain minerals exist and others not. The general table of inorganic structure types (Lima-de-Faria and Figueiredo, 1976, 1978) the layer description (Tables 1L to 17L) and the Tables 23S and 26S will certainly help to develop this problem.

Table 1S

A

CLOSE-PACKED

STRUCTURE TYPE	POPULATION	DERIVATIVES
OSMIUM $[Os]^{h}$ $P6_3/mmc$	IRIDOSMINE $[Os,Ir]^{h}$ SCHACHNERITE $[Ag_{1.1},Hg_{0.9}]^{h}$ RUTHENIRIDOSMINE $[Os,Ir,Ru]^{h}$ RUTHENIUM $[Ru]^{h}$ SILVER-2H $[Ag]^{h}$	CADMIUM $[Cd]^{h}$ c/a=1.88 ZINC $[Zn]^{h}$ c/a=1.86 (c/a ideal =1.63)
COPPER $[Cu]^{c}$ $Fm\bar{3}m$	SILVER-3C $[Ag]^{c}$ PLATINIRIDIUM $[Ir,Pt]^{c}$ GOLD $[Au]^{c}$ RHODIUM $[Rh]^{c}$ ELECTRUM $[Au,Ag]^{c}$ RUTHENOSMIRIDIUM $[Ir,Os,Ru]^{c}$ LEAD $[Pb]^{c}$ NICKEL $[Ni]^{c}$ PLATINUM $[Pt]^{c}$ ALUMINIUM $[Al]^{c}$ IRIDIUM $[Ir]^{c}$ PALLADIUM $[Pd]^{c}$	INDIUM $[In]^{c}$ $I\,4/mmm$ MERCURY $[Hg]^{c}$ $R\bar{3}m$ ($T< -39°C$)
IRON $[Fe]^{b}$ $Im\bar{3}m$	CHROMIUM $[Cr]^{b}$ MOSCHELLANDSBERGITE $[Ag_2,Hg_3]^{b}$	

GROUP

SULPHUR $\{g\}[S_8]$ Fddd
ROSICKYITE $\{g\}[S_8]^{\sim h}$ P2/c

CHAIN

SELENIUM $1\infty[Se^{[2]}]$ $P3_121$... ________________ Pop.: TELLURIUM $1\infty[Te^{[2]}]$

SHEET

GRAPHITE-2H $2\infty[C^{tr}]^{H\,h}$ $P6_3/mmc$
GRAPHITE-3R $2\infty[C^{tr}]^{H\,c}$ $R\bar{3}m$
ARSENOLAMPRITE $2\infty[As^{[3n]}]$ Bmab (Dist.deriv.Arsenic)
ARSENIC $2\infty[As^{[3n]}]^{\sim H\,c}$ $R\bar{3}m$ ________________

Pop.: BISMUTH $2\infty[Bi^{[3n]}]^{\sim H\,c}$
ANTIMONY $2\infty\,[Sb^{[3n]}]^{\sim H\,c}$
STIBARSEN $2\infty[Sb,As^{[3n]}]^{\sim H\,c}$
Deriv.: DURANUSITE $2\infty[As_4,S]^{\sim H\,c}$ Orth.

FRAMEWORK

LONSDALEITE $3\infty[C^{t}]^{\sim H\,h}$ $P6_3/mmc$
DIAMOND $3\infty[C^{t}]^{\sim H\,c}$ $Fd\bar{3}m$ ________________ Deriv.: TIN $3\infty[Sn^{t}]^{\sim H\,c}$ $I\,4_1/amd$

MINERALS TENTATIVELY CLASSIFIED

ALLARGENTUM $[Ag_{0.86},Sb_{0.14}]^{h}$ Hex.
CHROMFERIDE $[Fe_{1.5},Cr_{0.2},\square_{0.3}]^{b}$ $Pm\bar{3}m$
FERCHROMIDE $[Cr_{1.5},Fe_{0.2},\square_{0.3}]^{b}$ $Pm\bar{3}m$
KAMACITE $[Fe,Ni]^{b}$ $Im\bar{3}m$
LEADAMALGAM $[Pb_{0.7},Hg_{0.3}]^{c}$ $I\,4/mmm$
OSMIRIDIUM $[Ir,Os]^{c}$ $Fm\bar{3}m$
PARASCHACHNERITE $[Ag_{1.2},Hg_{0.8}]^{h}$ Cmcm
SILVER-4H $[Ag]^{ch}$ Hex.
TAENITE $[Ni,Fe]^{c}$ $Fm\bar{3}m$

MINERALS NOT YET CLASSIFIED

CHAOITE C P6/mmm

Table 2S

AB

CLOSE-PACKED

STRUCTURE TYPE	POPULATION	DERIVATIVES
TETRAAURICUPRIDE (Cuproaurite) $[CuAu]^{c}$ P4/mmm (Subs. deriv. Copper)	POTARITE $[PdHg]^{c}$ TETRATAENITE $[FeNi]^{c}$ TETRAFERROPLATINUM $[PtFe]^{c}$	
HONGSHIITE $[PtCu]^{c}$ R32... (Subs. deriv. Copper)		
WAIRAUITE $[CoFe]^{b}$ $Pm\bar{3}m$ (Subs. deriv. Iron)		
CAROBBIITE $[KF]^{Qs}$ $Fm\bar{3}m$		
NICKELINE (Niccolite) $Ni^{o}[As]^{h}$ $P6_3/mmc$	PYRRHOTITE-7C $(Fe_{1-x},\square_x)^{o}[S]^{h}$ BREITHAUPTITE $Ni^{o}[Sb]^{h}$ FREBOLDITE $Co^{o}[Se]^{h}$ SUDBURYITE $(Pd,Ni)^{o}[Sb]^{h}$ SEDERHOLMITE $Ni^{o}[Se]^{h}$ STUMPFLITE $Pt^{o}[Sb,Bi]^{h}$ KOTULSKITE $Pd^{o}[Te,Bi]^{h}$ NIGGLIITE $Pt^{o}[Sn]^{h}$ LANGISITE $(Co,Ni)^{o}[As]^{h}$	TROILITE $Fe^{o}[S]^{h}$ $P\bar{6}2c$ SMYTHITE $((Fe,Ni)_9\square_2)^{o}[S_{11}]^{h}$ $R\bar{3}m$ PLUMBOPALLADINITE $(Pb_2,\square)^{o}[Pd_3]^{h}$ Hex. PYRRHOTITE-4C $(Fe_7,\square)^{o}[S_8]^{h}$ C2/c... PYRRHOTITE-5C $(Fe_9,\square)^{o}[S_{10}]^{h}$ Hex. PYRRHOTITE-6C $(Fe_{11},\square)^{o}[S_{12}]^{h}$ Fd... PYRRHOTITE-11C $(Fe_{10},\square]^{o}[S_{11}]^{h}$ Hex. MODDERITE $(Co,Fe)^{o}[As]^{h}$ Pnma RUTHENARSENITE $(Ru,Ni)^{o}[As]^{h}$ Pnma WESTERVELDITE $(Fe,Ni,Co)^{o}[As]^{h}$ Pmcn
HALITE $Na^{o}[Cl]^{c}$ $Fm\bar{3}m$	CARLSBERGITE $Cr^{o}[N]^{c}$ SYLVITE $K^{o}[Cl]^{c}$ OSBORNITE $Ti^{o}[N]^{c}$ CHLORARGYRITE $Ag^{o}[Cl]^{c}$ ALABANDITE $Mn^{o}[S]^{c}$ BROMARGYRITE $Ag^{o}[Br]^{c}$ GALENA $Pb^{o}[S]^{c}$ BUNSENITE $Ni^{o}[O]^{c}$ NININGERITE $(Mg,Fe,Mn)^{o}[S]^{c}$ LIME $Ca^{o}[O]^{c}$ OLDHAMITE $(Ca,Mg)^{o}[S]^{c}$ MANGANOSITE $Mn^{o}[O]^{c}$ CLAUSTHALITE $Pb^{o}[Se]^{c}$ MONTEPONITE $Cd^{o}[O]^{c}$ ALTAITE $Pb^{o}[Te]^{c}$ PERICLASE $Mg^{o}[O]^{c}$ GRICEITE $Li^{o}[F]^{c}$ WÜSTITE $Fe^{o}[O]^{c}$ VILLIAUMITE $Na^{o}[F]^{c}$ TANTALCARBIDE $C^{o}[Ta]^{c}$	HERZENBERGITE $Sn^{o}[S]^{c}$ Pnma TEALLITE $(Pb,Sn)_2{}^{o}[S_2]^{c}$ Pmcn TRECHMANNITE $(Ag_{18},As_{18},\square_3)^{o}[S_{36}\square_3]^{c}$ $R\bar{3}$ SCHAPBACHITE $(Ag,Bi)_2{}^{o}[S_2]^{c}$ Cub. JORDANITE $(Pb_{14},(As,Sb)_6,\square_3)^{o}[S_{23}]^{c}$ $P2_1/m$
WURTZITE-2H $Zn^{t}[S]^{h}$ $P6_3mc$	GREENOCKITE $Cd^{t}[S]^{h}$ BROMELLITE $Be^{t}[O]^{h}$ CADMOSELITE $Cd^{t}[Se]^{h}$ ZINCITE $(Zn,Mn)^{t}[O]^{h}$ IODARGYRITE $Ag^{t}[I]^{h}$	
SPHALERITE (Blende) $Zn^{t}[S]^{c}$ $F\bar{4}3m$	BETA-MOISSANITE $Si^{t}[C]^{c}$ COLORADOITE $Hg^{t}[Te]^{c}$ HAWLEYITE $Cd^{t}[S]^{c}$ NANTOKITE $Cu^{t}[Cl]^{c}$ METACINNABAR $Hg^{t}[S]^{c}$ MARSHITE $Cu^{t}[I]^{c}$ STILLEITE $Zn^{t}[Se]^{c}$ MIERSITE $(Ag,Cu)^{t}[I]^{c}$ TIEMANNITE $Hg^{t}[Se]^{c}$	POLHEMUSITE $(Zn,Hg)^{t}[S]^{c}$ P4/n... MATRAITE $Zn^{t}[S]^{c}$ R3m. NUKUNDAMITE $(Cu_{3.4},Fe_{0.6})^{t}[S_4]^{c}$ $P\bar{3}m1$
LITHARGE $O^{t}[Pb]^{c}$ P4/nmm		VULCANITE $Cu^{t}[Te]^{c}$ Pmmn MASSICOT $O^{t}[Pb]^{c}$ Pbcn
SAL AMMONIAC $(NH_4)^{cb}[Cl]^{Qs}$ $Pm\bar{3}m$		ZHANGHENGITE $Cu^{cb}[Zn]^{Qs}$ $Pm\bar{3}m$
COOPERITE $(Pt,Pd,Ni)^{sq}[S]^{Qs}$ $P4_2/mmc$		TENORITE $Cu^{sq}[O]^{Qs}$ C2/c
VYSOTSKITE $(Pd,Ni)^{sq}[S]^{N21\,f}$ $P4_2/m$	BRAGGITE $(Pt,Pd,Ni)^{sq}[S]^{N21\,f}$	

Table 3S

AB(cont.)

GROUP

REALGAR $\{g\}[As_4^{[3]}S_4]$ $P2_1/n$
PARAREALGAR $\{g\}[As_4^{[3]}S_4]$ $P2_1/c$

CHAIN

MONTROYDITE $1\infty[Hg^{[2]}O]$ Pnma
CINNABAR $1\infty[Hg^{o}S^{«Tv»}]$ $P3_12_1$
CALOMEL $1\infty[Hg^{[5]}Cl^{«(2Qs)f»}]$ I 4/mmm ________________ Pop.: KUZMINITE $1\infty[Hg^{[5]}(Br,Cl)^{«(2Qs)f»}]$

FRAMEWORK

MILLERITE $3\infty[Ni^{[5]}S]$ R3m ________________________ Pop.: MÄKINENITE $3\infty[Ni^{[5]}Se]$

MINERALS TENTATIVELY CLASSIFIED

CHEREPANOVITE $Rh^{o}[As]^{h}$ Pnma
KHAMRABAEVITE $(Ti,V,Fe)^{o}[C]^{c}$ $Fm\bar{3}m$
MOISSANITE-5H $C^{t}[Si]^{c/h}$ Hex.
MOISSANITE-6H $C^{t}[Si]^{cch}$ $P6_3mc$

NEVSKITE $Bi^{o}[Se,S]^{c}$ $P\bar{3}m1$...
POLARITE $[Pd(Pb,Bi)]^{c}$ $Ccm2_1$
STISTAITE $Sn^{t}[Sb]^{c}$ Trig.
TSUMOITE $[BiTe]^{c}$ $P\bar{3}m1$

MINERALS NOT YET CLASSIFIED

CUPALITE (Cu,Zn)Al Orth.
EMPRESSITE AgTe Pmnm...
HYPERCINNABAR HgS Hex.
INGODITE Bi(Te,S) $P\bar{3}m1$

REALGAR (high) AsS α-form Mon.?
ROMARCHITE SnO P4/nmm
SOBOLEVSKITE PdBi $P6_3/mmc$

Table 4S

AB_2

CLOSE-PACKED

STRUCTURE TYPE	POPULATION	DERIVATIVES
CHALCOCITE $Cu_2^{tr/t}[S]^{h}$ $P2_1/c$	DJURLEITE $(Cu_{31},\square)^{tr/t}[S_{16}]^{h}$	
DIGENITE $Cu_2^{t/tr}[S]^{c}$ or $(Cu_9,\square)^{t/tr}[S_5]^{c}$ $R\bar{3}m$		BERZELIANITE $(Cu_{1.8},\square_{0.2})^{t/tr}[Se]^{c}$ $Fm\bar{3}m$ UMANGITE $(Cu_3,\square)^{t}[Se_2]^{c}$ $P\bar{4}2_1m$ ANILITE $(Cu_7,\square)^{t/tr}[S_4]^{c}$ Pnma ATHABASCAITE $(Cu_5,\square_3)^{t}[Se_4]^{c}$ Orth. GEERITE $(Cu_8,\square_2)^{t/tr}[S_5]^{c}$ $F\bar{4}3m$
MALDONITE $[[Au_2][Bi]]^{(.\sim TK)c}$ $Fd\bar{3}m$		
RUTILE $Ti^{o}[O_2]^{h}$ $P4_2/mnm$	SELLAITE $Mg^{o}[F_2]^{h}$ CASSITERITE $Sn^{o}[O_2]^{h}$ PLATTNERITE $Pb^{o}[O_2]^{h}$ PYROLUSITE $Mn^{o}[O_2]^{h}$ STISHOVITE $Si^{o}[O_2]^{h}$	TUGARINOVITE $Mo^{o}[O_2]^{h}$ $P2_1/c$
SCRUTINYITE $Pb^{o}[O_2]^{h}$ Pcan	IXIOLITE $(Ta,Mn,Nb)^{o}[O_2]^{h}$ SRILANKITE $(Ti,Zr)^{o}[O_2]^{h}$ CERVANTITE $Sb_2^{o}[O_4]^{h}$	
ANATASE $Ti^{o}[O_2]^{c}$ $I4_1/amd$		
BROOKITE $Ti^{o}[O_2]^{ch}$ Pbca		
RAMSDELLITE $Mn^{o}[O_2]^{h}$ Pbnm	PARAMONTROSEITE $V^{o}[O_2]^{h}$	
MELONITE $Ni^{o}[Te_2]^{h}$ $P\bar{3}m1$	BERNDTITE-2T $Sn^{o}[S_2]^{h}$ KITKAITE $Ni^{o}[(Te,Se)_2]^{h}$ MERENSKYITE $(Pd,Pt)^{o}[(Te,Bi)_2]^{h}$ MONCHEITE $(Pt,Pd)^{o}[(Te,Bi)_2]^{h}$ BRUCITE $Mg^{o}[(OH)_2]^{h}$ PORTLANDITE $Ca^{o}[(OH)_2]^{h}$ PYROCHROITE $Mn^{o}[(OH)_2]^{h}$ THEOPHRASTITE $Ni^{o}[(OH)_2]^{h}$	CALAVERITE $Au^{[6]}[Te_2]^{h}$ C2/m KRENNERITE $Au^{o}[Te_2]^{h}$ Pma2 CARLINITE $S^{o}[Tl_2]^{h}$ R3
CHLOROMAGNESITE $Mg^{o}[Cl_2]^{c}$ $R\bar{3}m$	SCACCHITE $Mn^{o}[Cl_2]^{c}$ LAWRENCITE $(Fe,Ni)^{o}[Cl_2]^{c}$	TOLBACHITE $Cu^{o}[Cl_2]^{c}$ C2/m
FLUORITE $Ca^{cb}[F_2]^{Qs}$ $Fm\bar{3}m$	FRANKDICKSONITE $Ba^{cb}[F_2]^{Qs}$ CERIANITE-(Ce) $Ce^{cb}[O_2]^{Qs}$ THORIANITE $Th^{cb}[O_2]^{Qs}$ URANINITE $U^{cb}[O_2]^{Qs}$	BADDELEYITE $Zr^{[7]}[O_2]^{Qs}$ $P2_1/c$
ANYUIITE $Au^{acb}[Pb_2]^{N21f}$ I 4mcm	KHATYRKITE $(Cu,Zn)^{acb}[Al_2]^{N21f}$	
MOLYBDENITE-2H $Mo^{p}[S_2]^{(2Ts)h}$ $P6_3/mmc$	TUNGSTENITE-2H $W^{p}[S_2]^{(2Ts)h}$ DRYSDALLITE $Mo^{p}[(Se,S)_2]^{(2Ts)h}$	
MOLYBDENITE-3R $Mo^{p}[S_2]^{(2Ts)c}$ $R\bar{3}m$	TUNGSTENITE-3R $W^{p}[S_2]^{(2Ts)c}$	

Table 5S

AB_2(cont.)

GROUP

STRUCTURE TYPE	POPULATION		DERIVATIVES
PYRITE $Fe^{o}\{g\}[S_2]^{c}$ $Pa\bar{3}$	CATTIERITE $Co^{o}\{g\}[S_2]^{c}$ ERLICHMANITE $Os^{o}\{g\}[S_2]^{c}$ VAESITE $Ni^{o}\{g\}[S_2]^{c}$ HAUERITE $Mn^{o}\{g\}[S_2]^{c}$ LAURITE $Ru^{o}\{g\}[S_2]^{c}$ PENROSEITE $Ni^{o}\{g\}[Se_2]^{c}$ KRUTAITE $Cu^{o}\{g\}[Se_2]^{c}$ TROGTALITE $Co^{o}\{g\}[Se_2]^{c}$ SPERRYLITE $Pt^{o}\{g\}[As_2]^{c}$	GERSDORFFITE-Pa3 $Ni^{o}\{g\}[(As,S)_2]^{c}$ HOLLINGWORTHITE $(Rh,Pt,Pd)^{o}\{g\}[(As,S)_2]^{c}$ IRARSITE $(Ir,Ru,Rh,Pt)^{o}\{g\}[(As,S)_2]^{c}$ PLATARSITE $(Pt,Rh,Ru)^{o}\{g\}[(As,S)_2]^{c}$ AUROSTIBITE $Au^{o}\{g\}[Sb_2]^{c}$ VILLAMANINITE $(Cu,Ni,Co,Fe)^{o}\{g\}[S_2]^{c}$ GEVERSITE $Pt^{o}\{g\}[Sb_2]^{c}$ MASLOVITE $Pt^{o}\{g\}[(Bi,Te)_2]^{c}$ INSIZWAITE $Pt^{o}\{g\}[(Bi,Sb)_2]^{c}$	KRUTOVITE $Ni^{o}\{g\}[As_2]^{c}$ $P2_13$
MARCASITE $Fe^{o}\{g\}[S_2^{«h»}]^{c}$ Pnnm	HASTITE $Co^{o}\{g\}[Se_2^{«h»}]^{c}$ KULLERUDITE $Ni^{o}\{g\}[Se_2^{«h»}]^{c}$ FERROSELITE $Fe^{o}\{g\}[Se_2^{«h»}]^{c}$ FROHBERGITE $Fe^{o}\{g\}[Te_2^{«h»}]^{c}$ MATTAGAMITE $(Co,Fe)^{o}\{g\}[Te_2^{«h«}]^{c}$ ANDUOITE $(Ru,Os)^{o}\{g\}[As_2^{«h»}]^{c}$	LÖLLINGITE $Fe^{o}\{g\}[As_2^{«h»}]^{c}$ RAMMELSBERGITE $Ni^{o}\{g\}[As_2^{«h»}]^{c}$ SAFFLORITE $Co^{o}\{g\}[As_2^{«h»}]^{c}$ SEINÄJOKITE $(Fe,Ni)^{o}\{g\}[(Sb,As)_2^{«h»}]^{c}$ NISBITE $Ni^{o}\{g\}[Sb_2^{«h»}]^{c}$	PARARAMMELSBERGITE $Ni^{o}\{g\}[As_2^{«h»}]^{c}$ Pbca

CHAIN

DOWNEYITE $1\infty[SeO_2]$ P4/mbc

SHEET

TELLURITE $2\infty[Te^{[4]}O_2]$ Pbcc

FRAMEWORK

CUPRITE $3\infty[Cu_2^{[2]}O^{t}]$ $Pn\bar{3}m$ ________________________ Deriv.: ACANTHITE $3\infty[Ag_2^{[2]}S^{t}]$ $P2_1/n$
AGUILARITE $3\infty[Ag_4^{[2]}(SeS)]$ Orth.
NAUMANNITE $3\infty[Ag_2^{[2]}Se^{t}]$ $P2_12_12_1$

QUARTZ $3\infty[Si^{t}O_2^{«Tv»}]$ $P3_121$
TRIDYMITE $3\infty[Si^{t}O_2^{[2]«ch»}]^{h}$ Cc
CRISTOBALITE $3\infty[Si^{t}O_2^{[2]«c»}]^{c}$ $P4_12_12$
COESITE $3\infty[Si^{t}O_2^{[2]}]^{c}$ C2/c
ICE $3\infty[O^{t}H_2^{[2]«ch»}]^{h}$ $P6_3/mmc$

MINERALS TENTATIVELY CLASSIFIED

BAMBOLLAITE $Cu^{t}[(Se,Te)_2]^{c}$ $P4_2/n$
BELLIDOITE $Cu_2^{t}\{Se\}^{c}$ $P4_2/n$
BERNDTITE-4H $Sn^{o}[S_2]^{c/h}$ $P6_3mc$
CLINOSAFFLORITE $(Co,Fe,Ni)^{o}\{g\}[As_2]$ $P2_1/n$
CUPROSTIBITE $Cu^{o}Cu^{t}[Sb,Tl]^{c}$ P4/nmm
HEDLEYITE $Bi^{o}[Bi_{1+x},Te_{1-x}]^{c/h}$ $R\bar{3}m$

IRIDARSENITE $(Ir,Ru)\{g\}[As_2^{«h»}]^{c}$ Mon.
NSUTITE $Mn^{o}[(O,OH)_2]^{h}$ Hex.
ROXBYITE $(Cu_{1.78},\square_{0.22})^{tr/t}[S]^{h}$ C2/m...
WÜLFINGITE $3\infty[Zn(OH)_2]$ $P2_12_12_1$
ZIRKELITE $(Ti,Ca,Zr)^{cb}[O_{2-x}]^{Qs}$ Cub.

MINEALS NOT YET CLASSIFIED

AKHTENSKITE ε-MnO_2 $P6_3/mmc$
AMAKINITE $(Fe,Mg)(OH)_2$ $R\bar{3}m$
ARGUTITE GeO_2 Tet.
ASHANITE $(Nb,Ta,Fe,Mn,V)_4O_8$ Pcan
ASHOVERITE $Zn(OH)_2$ $I4_1/amd$...
●**BARRINGERITE** $(Fe,Ni)_2P$ P62m
BEHOITE β-$Be(OH)_2$ $P2_12_12_1$
BORISHANSKIITE $Pd(As,Pb)_2$ $Ccm2_1$
●**CLINOBEHOITE** $Be(OH)_2$ $P2_1$
COTUNNITE $PbCl_2$ Pbmn
DANBAITE $CuZn_2$ Cub.

FROODITE α-$PdBi_2$ C2/m
LAURELITE $Pb(F,Cl,OH)_2$ P6
MCKINSTRYITE $(Ag,Cu)_2S$ Pnam...
PALLADOARSENIDE Pd_2As P2/m
PALLADOBISMUTHARSENIDE $Pd_2(As,Bi)$ Pmcn...
PAOLOVITE Pd_2Sn Pbnm
PARATELLURITE TeO_2 $P4_12_12$
PAXITE $CuAs_2$ $P2_1/c$
SWEETITE $Zn(OH)_2$ $P4_12_12$...
URVANTSEVITE $Pd(Bi,Pb)_2$ Hex.
WEISSITE $Cu_{2-x}Te$ P6/nmm

● The structure has been determined, but we could not interpreted it.

Table 6S

AB_3

CLOSE-PACKED

STRUCTURE TYPE	POPULATION	DERIVATIVES
DYSCRASITE $[Ag_3Sb]^h$ Pmm2 (subs. deriv.Osmium)		
AURICUPRIDE $[Cu_3Au]^c$ $Pm\bar{3}m$ (subs. deriv. Copper)	ZVYAGINTSEVITE $[Pd_3(Pb,Sn)]^c$ AWARUITE $[Ni_3Fe]^c$	
MOLYSITE $Fe^O[Cl_3]^h$ $R\bar{3}$		
MOLYBDITE $Mo^o[O_3]^c$ Pbnm		
GIBBSITE $Al^o[(OH)_3]^{(2Ts)h}$ $P2_1/n$	BAYERITE $Al^o[(OH)_3]^{(2Ts)h}$	
DZHALINDITE $In^o[\square(OH)_3]^c$ $I\ m\bar{3}$ (defect deriv. Perovskite)	SÖHNGEITE $Ga^o[\square(OH)_3]^c$	

GROUP

SKUTTERUDITE $Co_4^o\{g\}[As_4]_3$ Im3

MINERALS TENTATIVELY CLASSIFIED

ATOKITE $[(Pd,Pt)_3Sn]^c$ $Fm\bar{3}m$
CABRIITE $[(Pd,Cu)_3Sn]^c$ Pnmc
COHENITE $C^p[Fe_3]^{Ts}$ Pnma
DOMEYKITE $Cu_3^t[As]^b$ $I\bar{4}3d$
DOYLEITE $Al^o[(OH)_3]^{c/h}$ P1
FLUOCERITE-(Ce) $(Ce,La)^{tr}[F_3]^{cch}$ $P6_3/mcm$
FLUOCERITE-(La) $(La,Ce)^{tr}[F_3]^{cch}$ P3c1
GANANITE $Bi^o[F_3]^c$ $P\bar{4}3m$
GUPEIITE $[Fe_3Si]^b$ $Fm\bar{3}m$
NICKEL-SKUTTERUDITE $Ni_4^o\{g\}[As_4]_{2\text{-}3}$ I m3
NORDSTRANDITE $Al^o[(OH)_3]^{(2Ts)c}$ P1...
RUSTENBURGITE $[Pt_3Sn]^c$ Fm3m
SUESSITE $[Fe_3Si]^b$ $Fm\bar{3}m$

MINERALS NOT YET CLASSIFIED

ATHENEITE $(Pd,Hg)_3As$ P6/mmm
ISOFERROPLATINUM $(Pt,Pd)_3(Fe,Cu)$ Pm3m...
LUANHEITE Ag_3Hg Hex.
PARADOCRASITE Sb_3As C2
SASSOLITE $B(OH)_3$ P1
●**SCHREIBERSITE** $(Fe,Ni,Cr)_3P$ I 4
SIDERAZOT $Fe_{2.5}N$ $P\bar{3}31m$
TELARGPALITE $(Pd,Ag)_3Te$ Cub.
VINCENTITE $(Pd,Pt)_3(As,Sb,Te)$ Tet.

Table 7S

A_2B_3

CLOSE-PACKED

STRUCTURE TYPE	POPULATION	DERIVATIVES
CORUNDUM $Al_2^{o}[O_3]^{h}$ $R\bar{3}c$	ESKOLAITE $Cr_2^{o}[O_3]^{h}$ HEMATITE $Fe_2^{o}[O_3]^{h}$ KARELIANITE $V_2^{o}[O_3]^{h}$	
TELLUROBISMUTHITE $Bi_2^{o}[Te_3]^{cch}$ $R\bar{3}m$	TELLURANTIMONY $Sb_2^{o}[Te_3]^{cch}$	
BIXBYITE $Mn_2^{[6]}[O_3\square]^{Qs}$ $Ia\bar{3}$ (defect deriv. Fluorite)	AVICENNITE $Tl_2^{[6]}[O_3\square]^{Qs}$	
STIBNITE $Sb_2^{[3+4]}[S_3]^{\#}$ Pbnm (non layered)	BISMUTHINITE $Bi_2^{[3+4]}[S_3]^{\#}$ GUANAJUATITE $Bi_2^{[3+4]}[Se_3]^{\#}$	

CHAIN

VALENTINITE $_{1\infty}[Sb_2^{[3]}O_3]$ Pccn

SHEET

ORPIMENT $_{2\infty}[As_2^{[3n]}S_3]$ $P2_1/n$ ________________ Pop. LAPHAMITE $_{2\infty}[As_2^{[3n]}(Se,S)_3]$

MINERALS TENTATIVELY CLASSIFIED

BOWIEITE $(Rh,Ir,Pt)_2^{o}[S_3]^{h}$ Pbcn
HEAZLEWOODITE $Ni_3^{t}[S_2]^{b}$ R32
KASHINITE $(Ir,Rh)_2^{o}[S_3]^{b}$ Pbcn
MOSCHELLANDSBERGITE $[Ag_2Hg_3]^{b}$ Im3m

MINERALS NOT YET CLASSIFIED

ARSENOLITE As_2O_3 $Fd\bar{3}m$
BISMITE α-Bi_2O_3 $P2_1/c$
●**CLAUDETITE** As_2O_3 $P2_1/n$
MONTBRAYITE $(Au,Sb)_2Te_3$ P1
●**OTTEMANNITE** Sn_2S_3 Pnma
PARAGUANAJUATITE $Bi_2(Se,S)_3$ $R\bar{3}m$
●**SENARMONTITE** Sb_2O_3 $Fd\bar{3}m$
STANNOPALLADINITE Pd_3Sn_2 $P6_3/mmc$
TAZHERANITE $(Zr,Ti,Ca)_4O_6$ $Fm\bar{3}m$
●**TONGBAITE** Cr_3C_2 Pnma
WEISHANITE $(Au,Ag)_3Hg_2$ $P6_3/mmc$

AB_4

CLOSE-PACKED

ROALDITE $N^{o}[(Fe,Ni)_4]^{c}$ $Pm\bar{3}m$

CHAIN

PATRÓNITE $_{1\infty}[V^{[8]}S_4]$ I 2/c

Table 8S

A_3B_4

MINERALS TENTATIVELY CLASSIFIED

DIMORPHITE-I {g}[As_4S_3] Pnma...
DIMORPHITE-II {g}[As_4S_3] Pnma...
IKUNOLITE $Bi_4^{o}[(S,Se)_3]^{c/h}$ R$\bar{3}$m
LAITAKARITE $Bi_4^{o}[(Se,S)_3]^{c/h}$ R$\bar{3}$m
RUCKLIDGEITE $(Bi,Pb)_3^{o}[Te_4]^{c/h}$ R$\bar{3}$m

MINERALS NOT YET CLASSIFIED

BREZINAITE Cr_3S_4 I 2/m
GENKINITE $(Pt,Pd)_4Sb_3$ P422
PILSENITE Bi_4Te_3 R$\bar{3}$m
WILKMANITE Ni_3Se_4 I 2/m

A_2B_5

MINERALS TENTATIVELY CLASSIFIED

SHCHERBINAITE $V_2^{[5+1]}[O_5]^{c}$ Pmmn

MINERALS NOT YET CLASSIFIED

KOUTEKITE Cu_5As_2 Hex.
ORCELITE $Ni_{5-x}As_2$ Hex.
PERRYITE $(Ni,Fe)_5(Si,P)_2$ Hex.
STIBIOPALLADINITE Pd_5Sb_2 P6_3/mmc
TANTITE Ta_2O_5 P1...

AB_6

MINERALS NOT YET CLASSIFIED

ALGODONITE Cu_6As P6_3/mmc

Table 9S

$A_mB_n(m \geq 3, n \geq 5)$

MINERALS TENTATIVELY CLASSIFIED

RICKARDITE $[Cu_7Te_5]^{(2Qs)f}$ P4/nmm

MINERALS NOT YET CLASSIFIED

ALACRANITE As_8S_9 P2/c
ARSENOPALLADINITE $Pd_8(Sb,As)_3$ P1...
FERRIHYDRITE $Fe_{4-5}(OH,O)_{12}$ Hex.
HAXONITE $(Fe,Ni)_{23}C_6$ Fm$\bar{3}$m
ISOMERTIEITE $Pd_{11}(Sb,As)_4$ Fd$\bar{3}$m
KEITHCONNITE $Pd_{20}Te_7$ R$\bar{3}$...
●**KOLYMITE** Cu_7Hg_6 I $\bar{4}$3m
MACKINAWITE $(Fe,Ni)_9S_8$ P4_2/nmm
●**MAUCHERITE** $Ni_{11}As_8$ P4_12_12
MERTIEITE-I $Pd_{11}(Sb,As)_4$ Hex.
MERTIEITE-II $Pd_8(Sb,As)_3$ Trig.
NOVÁKITE $(Cu,Ag)_{21}As_{10}$ C2
OOSTERBOSCHITE $(Pd,Cu)_7Se_5$ Orth.
PALLADSEITE $Pd_{17}Se_{15}$ Pm$\bar{3}$m
SPIONKOPITE $Cu_{39}S_{28}$ P$\bar{3}$m1
STILLWATERITE Pd_8As_3 P$\bar{3}$...
STÜTZITE $Ag_{5-x}Te_3$ P6/mmm
TELLUROPALLADINITE Pd_9Te_4 P2_1/c
UZONITE As_4S_5 P2_1/m
XIFENGITE Fe_5Si_3 P6_3/mmc
YARROWITE Cu_9S_8 P$\bar{3}$m1

Table 10S

ABC

CLOSE-PACKED

STRUCTURE TYPE	POPULATION	DERIVATIVES
MANGANITE $Mn^{ol}[O(OH)]^{h}$ $B2_1/d$ (subs.deriv. Rutile)		GUYANAITE $Cr^{o}[O(OH)]^{h}$ Pnnm
DIASPORE $Al^{o}[O(OH)]^{h}$ Pbnm (subs.deriv. Ramsdellite)	BRACEWELLITE $Cr^{o}[O(OH)]^{h}$ GOETHITE $Fe^{o}[O(OH)]^{h}$ GROUTITE $Mn^{o}[O(OH)]^{h}$ MONTROSEITE $(V,Fe)^{o}[O(OH)]^{h}$	
HETEROGENITE-2H $Co^{o}[O(OH)]^{(2Ts)h}$ $P6_3/mmc$		
HETEROGENITE-3R $Co^{o}[O(OH)]^{(2Ts)c}$ $R\bar{3}m$	GRIMALDIITE $Cr^{o}[O(OH)]^{(2Ts)c}$	
MATLOCKITE $Pb^{[9]}[[F][Cl]]^{(Q2,2Q1\,f)\,f}$ P4/nmm	BISMOCLITE $Bi^{[9]}[[Cl][O]]^{(Q2,2Q1\,f)\,f}$ DAUBRÉEITE $Bi^{[9]}[[Cl][OH]]^{(Q2,2Q1\,f)\,f}$ ZAVARITSKITE $Bi^{[9]}[[F][O]]^{(Q2,2Q1\,f)\,f}$	

GROUP

STRUCTURE TYPE	POPULATION	DERIVATIVES
ULLMANNITE $Ni^{o}\{g\}[SbS]^{c}$ $P2_13$ (subs. deriv. Pyrite)	GERSDORFFITE-$P2_13$ $Ni^{o}\{g\}[AsS]^{c}$ MICHENERITE $(Pd,Pt)^{o}\{g\}[BiTe]^{c}$ WILLYAMITE $(Co,Ni)^{o}\{g\}[SbS]^{c}$	
COBALTITE $Co^{o}\{g\}[AsS]^{c}$ $Pca2_1$ (subs.deriv, Pyrite)	GERSDORFFITE-$Pca2_1$ $Ni^{o}\{g\}[AsS]^{c}$ TOLOVKITE $Ir^{o}\{g\}[SbS]^{c}$	ALLOCLASITE $(Co,Fe)^{o}\{g\}[AsS]^{c}$ $P2_1$
ARSENOPYRITE $Fe^{o}\{g\}[AsS^{*sh*}]^{c}$ $B2_1/d$ (subs. deriv. Marcasite)	GUDMUNDITE $Fe^{o}\{g\}[SbS^{*sh*}]^{c}$ OSARSITE $(Os,Ru)^{o}\{g\}[AsS^{*sh*}]^{c}$	COSTIBITE $Co^{o}\{g\}[SbS^{*sh*}]^{c}$ $Pn2_1m$ GLAUCODOT $(Co,Fe)^{o}\{g\}[AsS^{*sh*}]^{c}$ Cmmn

MINERALS TENTATIVELY CLASSIFIED

BÖHMITE $Al^{o}[O(OH)]^{R31\,my}$ Bbmn
EUCAIRITE $Cu^{t}Ag^{t}[Se]^{c}$ Orth.
LAURIONITE $Pb^{[9]}[[Cl][OH]]^{(Q2,2Q1\,f)\,f}$ Pcmn
LAUTITE $Cu^{t}[AsS]^{c}$ $Pna2_1$
LEPIDOCROCITE $Fe^{o}[O(OH)]^{R31\,my}$ $Cmc2_1$

HESSITE $Ag^{[5]}Ag^{t}[Te]^{c}$ $P2_1/c$
PARACOSTIBITE $Co\{g\}[SbS]^{c}$ Pbca
PARALAURIONITE $Pb^{[9]}[[Cl][OH]]^{(Q2,2Q1\,f)\,f}$ C2/m
SPERTINIITE $Cu[(OH)_2]^{R31\,my}$ Bbmn
STROMEYERITE $Cu^{t/tr}Ag^{t/tr}[S]^{c}$ Cmcm

MINERALS NOT YET CLASSIFIED

FEITKNECHTITE β-MnO(OH) Tet.
FEROXYHYTE δ-FeO(OH) Hex.

MAJAKITE PdNiAs P62m
PETROVSKAITE AuAg(S,Se) P2m...

Table 11S

ABC_2

CLOSE-PACKED

STRUCTURE TYPE	POPULATION	DERIVATIVES
MATILDITE $Ag^{o}Bi^{o}[S_2]^{c}$ $P\bar{3}m1$ (subs.deriv. Halite)		
CHALCOPYRITE $Cu^{t}Fe^{t}[S_2]^{c}$ $I\bar{4}2d$ (subs.deriv. Sphalerite)	GALLITE $Cu^{t}Ga^{t}[S_2]^{c}$ ROQUESITE $Cu^{t}In^{t}[S_2]^{c}$	RAGUINITE $Tl^{t}Fe^{t}[S_2]^{c}$ Orth.

MINERALS TENTATIVELY CLASSIFIED

CASWELLSILVERITE $Na^{o}[Cr^{[2]}S_2]^{(3Ts)\,c}$ $R\bar{3}m$
CHALCOSTIBITE $Cu^{t}Sb^{t}[S_2]^{h}$ Pnam
DELAFOSSITE $Fe^{o}[Cu^{[2]}O_2]^{(3Ts)\,c}$ $R\bar{3}m$
EMPLECTITE $Cu^{t}Bi^{[3n]}[S_2]^{h}$ Pnma
MCCONNELLITE $Cr^{o}[Cu^{[2]}O_2]^{(3Ts)\,c}$ $R\bar{3}m$
METABORITE $3\infty[B^{t}O_2H]$ $P\bar{4}3n$
MIARGYRITE $Ag^{[2/3]}Sb^{[3n]}[S_2]^{c}$ Cc
PUTORANITE $Cu_{1.1}{}^{t}Fe_{1.2}{}^{t}[S_2]^{c}$ Cub.
SMITHITE $Ag^{[3]}As^{[2]}[S_2]^{c}$ C2/c
TERLINGUAITE $1\infty[Hg_2OCl]$ C2/c

MINERALS NOT YET CLASSIFIED

● **ARAMAYOITE** $Ag(Sb,Bi)S_2$ P1
BOHDANOWICZITE $AgBiSe_2$ $P\bar{3}m1$
CREDNERITE $CuMnO_2$ C2m
● **DUTTONITE** $VO(OH)_2$ I 2/c
FERRONICKELPLATINUM Pt_2FeNi Tet.
KOLARITE $PbTeCl_2$ Pmmm...
● **LORANDITE** $TlAsS_2$ $P2_1/a$
MUTHMANNITE $AuAgTe_2$ (?)
TULAMEENITE Pt_2FeCu Tet.
VOLYNSKITE $AgBiTe_2$ $R\bar{3}m$
WEISSBERGITE $TlSbS_2$ P1
ZHARCHIKHITE $AlF(OH)_2$ $P2_1/a$

Table 12S

ABC_3

CLOSE-PACKED

STRUCTURE TYPE	POPULATION	DERIVATIVES
ILMENITE $Fe^{o}Ti^{o}[O_3]^{h}$ $R\bar{3}$ (subs.deriv. Corundum)	ECANDREWSITE $(Zn,Fe,Mn)^{o}Ti^{o}[O_3]^{h}$ GEIKIELITE $Mg^{o}Ti^{o}[O_3]^{h}$ MELANOSTIBITE $Mn^{o}(Sb,Fe)^{o}[O_3]^{h}$ PYROPHANITE $Mn^{o}Ti^{o}[O_3]^{h}$	
PEROVSKITE $Ti^{o}[CaO_3]^{c}$ Pcmn (subs.deriv. Roaldite)	LATRAPPITE $(Nb,Ti)^{o}[(Ca,Na)O_3]^{c}$ NEIGHBORITE $Mg^{o}[NaF_3]^{c}$	LOPARITE-(Ce) $(Ti,Nb)^{o}[(Ce,Na,Ca)O_3]^{c}$ Pm3m LUESHITE $Nb^{o}[NaO_3]^{c}$ $P222_1$ MACEDONITE $Ti^{o}[PbO_3]^{c}$ P4/mmm TAUSONITE $Ti^{o}[SrO_3]^{c}$ Pm3m

GROUP

STRUTURE TYPE	POPULATION	
ARAGONITE $Ca^{[6/9]}\{g\}[C^{tr}O_3]$ Pmcn	NITER $K^{[6/9]}\{g\}[N^{tr}O_3]$ CERUSSITE $Pb^{[6/9]}\{g\}[C^{tr}O_3]$ STRONTIANITE $Sr^{[6/9]}\{g\}[C^{tr}O_3]$ WITHERITE $Ba^{[6/9]}\{g\}[C^{tr}O_3]$	
CALCITE $Ca^{ap}\{g\}[C^{tr}O_3{}^{«h»}]^{c}$ $R\bar{3}c$	NITRATINE $Na^{ap}\{g\}[N^{tr}O_3{}^{«h»}]^{c}$ GASPÉITE $(Ni,Mg,Fe)^{ap}\{g\}[C^{tr}O_3{}^{«h»}]^{c}$ MAGNESITE $Mg^{ap}\{g\}[C^{tr}O_3{}^{«h»}]^{cc}$ OTAVITE $Cd^{ap}\{g\}[C^{tr}O_3{}^{«h»}]^{c}$	RHODOCHROSITE $Mn^{ap}\{g\}[C^{tr}O_3{}^{«h»}]^{c}$ SIDERITE $Fe^{ap}\{g\}[C^{tr}O_3{}^{«h»}]^{c}$ SMITHSONITE $Zn^{ap}\{g\}[C^{tr}O_3{}^{«h»}]^{c}$ SPHAEROCOBALTITE $Co^{ap}\{g\}[C^{tr}O_3{}^{«h»}]^{c}$
VATERITE $Ca^{[6]}\{g\}[C^{tr}O_3]^{h}$ $P6_3/mmc$...		

CHAIN

PYROXFERROITE $(Fe,Mn,Ca)_7{}^{o}\{1\infty\}[Si_7{}^{t}O_{21}]$ P1...
PYROXMANGITE $Mn_7{}^{o}\{1\infty\}[Si_7{}^{t}O_{21}]$ P1
RHODONITE$(Mn,Fe,Mg,Ca)_5{}^{[7]}\{1\infty\}[Si_5{}^{t}O_{15}]$ P1
WOLLASTONITE-1T $Ca^{o}\{1\infty\}[Si^{t}O_3]$ P1
WOLLASTONITE-2M $Ca^{o}\{1\infty\}[Si^{t}O_3]$ $P2_1/a$
WOLLASTONITE-3T $Ca^{o}\{1\infty\}[Si^{t}O_3]$ Tric.
WOLLASTONITE-4T $Ca^{o}\{1\infty\}[Si^{t}O_3]$ Tric.
WOLLASTONITE-5T $Ca^{o}\{1\infty\}[Si^{t}O_3]$ Tric.
WOLLASTONITE-7T $Ca^{o}\{1\infty\}[Si^{t}O_3]$ Tric.

SHEET

GETCHELLITE $2\infty[Sb^{[3n]}As^{[3n]}S_3{}^{[2]}]$ $P2_1/a$

MINERALS TENTATIVELY CLASSIFIED

ALAMOSITE $Pb^{[3/4]}\{1\infty\}[Si^{t}O_3]$ P2/n
KANOITE $(Mn,Mg)_2{}^{[6]}\{1\infty\}[Si_2{}^{t}O_6]$ $P2_1/c$
KIMROBINSONITE $Ta^{o}[(O,CO_3)(OH)_3]^{c}$ Cub
NITRAMMITE $(NH_4)^{[8+4]}\{g\}[N^{tr}O_3]$ Pmm

MINERALS NOT YET CLASSIFIED

BALYAKINITE $CuTeO_3$ Pmcn
CHLOROCALCITE $KCaCl_3$ Pnma
FAIRBANKITE $PbTeO_3$ $P\bar{1}$
● **MOLYBDOMENITE** $PbSeO_3$ $P2_1/m$
PLUMBOTELLURITE α-$PbTeO_3$ Orth.
POYARKOVITE Hg_3OCl C2/m...
SCOTLANDITE $PbSO_3$ $P2_1/m$...

TABLE 13S

ABC_4

CLOSE-PACKED

STRUCTURE TYPE	POPULATION	DERIVATIVES
WOLFRAMITE $(Fe,Mn)^{o}W^{o}[O_4]^{h}$ P2/c (subs.deriv.Scrutinyite)	FERBERITE $Fe^{o}W^{o}[O_4]^{h}$ HÜBNERITE $Mn^{o}W^{o}[O_4]^{h}$ SANMARTINITE $(Zn,Fe)^{o}W^{o}[O_4]^{h}$	BISMUTOTANTALITE $Bi^{o}(Ta,Nb)^{o}[O_4]^{h}$ Pcmn STIBIOTANTALITE $Sb^{o}Ta^{o}[O_4]^{h}$ $Pc2_1/n$ STIBIOCOLUMBITE $Sb^{o}Nb^{[6]}[O_4]^{h}$ $Pc2_1n$
HETEROSITE $(Fe,Mn)^{o}P^{t}[O_4]^{h}$ Pmnb (relat. Forsterite)	PURPURITE $(Mn,Fe)^{o}P^{t}[O_4]^{h}$	
ANHYDRITE $Ca^{[8]}S^{t}[O_4]^{*}$ Bbmn (non layered)	FERRUCCITE $Na^{[8]}B^{t}[F_4]^{*}$	
SCHEELITE $Ca^{do}W^{o}[O_4]^{*}$ $I\,4_1/a$ (non layered)	FERGUSONITE-(Y) $Y^{do}Nb^{t}[O_4]^{*}$ FORMANITE-(Y) $Y^{do}Ta^{t}[O_4]^{*}$ POWELLITE $Ca^{do}Mo^{t}[O_4]^{*}$ STOLZITE $Pb^{do}W^{t}[O_4]^{*}$ WULFENITE $Pb^{do}Mo^{t}[O_4]^{*}$	BETA-FERGUSONITE-(Y) $Y^{do}Nb^{t}[O_4]^{*}$ C2/c
ZIRCON $Zr^{do}Si^{t}[O_4]^{*}$ $I\,4_1/amd$ (non layered)	BEHIERITE $(Ta,Nb)^{do}B^{t}[O_4]^{*}$ CHERNOVITE-(Y) $Y^{do}As^{t}[O_4]^{*}$ CHROMATITE $Ca^{do}Cr^{t}[O_4]^{*}$ DREYERITE $Bi^{do}V^{t}[O_4]^{*}$ HAFNON $Hf^{do}Si^{t}[O_4]^{*}$ THORITE $Th^{do}Si^{t}[O_4]^{*}$ XENOTIME-(Y) $Y^{do}P^{t}[O_4]^{*}$ WAKEFIELDITE-(Ce) $(Ce,Pb)^{do}V^{t}[O_4]^{*}$ WAKEFIELDITE-(Y) $Y^{do}V^{t}[O_4]^{*}$	
MONAZITE-(Ce) $(Ce,La,Nd,Th)^{[9]}P^{t}[O_4]^{*}$ $P2_1/n$ (non layered)	CHERALITE $(Th,Ca,Ce)^{[9]}(P,Si)^{t}[O_4]^{*}$ CROCOITE $Pb^{[9]}Cr^{t}[O_4]^{*}$ HUTTONITE $Th^{[9]}Si^{t}[O_4]^{*}$ MONAZITE-(La) $(La,Ce,Nd)^{[9]}P^{t}[O_4]^{*}$ MONAZITE-(Nd) $(Nd,La,Ce)^{[9]}P^{t}[O_4]^{*}$ ROOSEVELTITE $Bi^{[9]}As^{t}[O_4]^{*}$	

GROUP

BARITE $Ba^{[12]}\{g\}[S^{t}O_4]$ Pbnmm ———————— Pop.: ANGLESITE $Pb^{[12]}\{g\}[S^{t}O_4]$
AVOGADRITE $(K,Cs)^{[12]}\{g\}[B^{t}F_4]$
HASHEMITE $Ba^{[12]}\{g\}[(Cr,S)^{t}O_4]$
CELESTINE $Sr^{[12]}\{g\}[S^{t}O_4]$

CHAIN

PUCHERITE $\{1\infty\}[Bi^{do}V^{t}O_4]$ Pnca

FRAMEWORK

BERLINITE $3\infty[Al^{t}P^{t}O_4]$ $P3_121$

MINERALS TENTATIVELY CLASSIFIED

BETA-FERGUSONITE-(Ce) $(Ce,La,Nd)^{do}Nb^{t}[O_4]^{*}$ Mon.
BETA-FERGUSONITE-(Nd) $(Nd,Ce)^{do}Nb^{t}[O_4]^{*}$ Mon.
CHALCOCYANITE $Cu^{o}\{g\}[S^{t}O_4]$ Pnma
CLINOBISVANITE $Bi^{do}V^{t}[O_4]^{*}$ I 2/a
RASPITE $Pb^{[7]}\{1\infty\}[W^{o}O_4]$ $P2_1/a$
SYLVANITE $Ag^{o}Au^{o}[Te_4]^{h}$ P2/c

MINERALS NOT YET CLASSIFIED

ALUMOTANTITE $AlTaO_4$ Orth.
BEZSMERTNOVITE $(Au,Ag)_4Cu(Te,Pb)$ Orth.
CERVELLEITE Ag_4TeS P23...
GASPARITE-(Ce) $(Ce,La,Nd)AsO_4$ $P2_1/n$
KOSTOVITE $CuAuTe_4$ Pma2

Table 14S

ABC_5

MINERALS NOT YET CLASSIFIED

PALARSTANIDE Pd_5SnAs Trig.

ABC_6

MINERALS TENTATIVELY CLASSIFIED

STOTTITE $Fe^{o}Ge^{o}[\square_2(OH)_6]^{c}$ $P4_2/n$ (subs.deriv. Dzhalindite)-- Pop.:TETRAWICKMANITE $Mn^{o}Sn^{o}[\square_2(OH)_6]^{c}$
Deriv.: WICKMANITE $Mn^{o}Sn^{o}[\square_2(OH)_6]^{c}$ $Pn\bar{3}m$

MINERALS NOT YET CLASSIFIED

BURTITE $CaSn(OH)_6$ R3
JEANBANDYITE $(Fe,Mn)Sn(OH)_6$ $P4_2/n$
MOPUNGITE $NaSb(OH)_6$ $P4_2/n$
MUSHISTONITE $(Cu,Zn,Fe)Sn(OH)_6$ Pn3m
NATANITE $FeSn(OH)_6$ $Pn\bar{3}m$
SCHOENFLIESITE $MgSn(OH)_6$ $Pn\bar{3}$
VISMIRNOVITE $ZnSn(OH)_6$ $Pn\bar{3}m$

AB_2C_2

CLOSE-PACKED

TETRADYMITE $Bi_2{}^{o}[Te_2S]^{chh}$ or $Bi_{14}{}^{o}[Te_{13}S_8]^{chh}$ $R\bar{3}$ Pop.: KAWAZULITE $Bi_2{}^{o}[Te_2Se]^{chh}$

MINERALS TENTATIVELY CLASSIFIED

IMITERITE $Ag_2{}^{tr}Hg^{o}[S_2]^{h}$ $P2_1/c$

MINERALS NOT YET CLASSIFIED

BLIXITE $Pb_2Cl(O,OH)_2$ Orth.
CHVILEVAITE $Na(Cu,Fe,Zn)_2S_2$ P3m1
DERVILLITE Ag_2AsS_2 P2/a
●**KERMESITE** Sb_2OS_2 P1
MELANOTHALLITE Cu_2OCl_2 Fddd
OREGONITE $FeNi_2As_2$ Hex.
PÄÄKKÖNENITE Sb_2AsS_2 Mon.
SINOITE Si_2N_2O Orth.
SKIPPENITE Bi_2Se_2Te $R\bar{3}m$

AB_2C_3

CLOSE-PACKED

CUBANITE $Cu^{t}Fe_2{}^{t}[S_3]^{h}$ Pcmn _______________ Pop.: ARGENTOPYRITE $Ag^{t}Fe_2{}^{t}[S_3]^{h}$
COVELLITE $Cu_2{}^{t}Cu^{tr}[S_3]^{(3Th)s}$ $P6_3/mmc$ ______________ Pop.: KLOCKMANNITE $Cu_2{}^{t}Cu^{tr}[Se_3]^{(3Th)s}$

MINERALS TENTATIVELY CLASSIFIED

ATACAMITE $Cu_2{}^{[6]}[Cl(OH)_3]^{c}$ Pnma
BOTALLACKITE $Cu_2{}^{[6]}[Cl(OH)_3]^{c}$ $P2_1/m$
FISCHESSERITE $Ag_3{}^{[4]}Au^{[2]}[Se_2{}^{[8]}]^{h}$ I 432
ISOCUBANITE $Cu^{t}Fe_2{}^{t}[S_3]^{c}$ $Fm\bar{3}m$
MOHITE $Cu_2{}^{t}Sn^{t}[S_3]^{h}$ P1
PETZITE $Ag_3{}^{[4]}Au^{[2]}[Te_2]^{c}$ $I\,4_132$
PICOTPAULITE $Tl^{t}Fe_2{}^{t}[S_3]^{h}$ C222...

MINERALS NOT YET CLASSIFIED

GREGORYITE α-$(Na,K,Ca)_2CO_3$ $P6_3mc$
JALPAITE Ag_3CuS_2 $I\,4_1/amd$
KEMPITE $Mn_2Cl(OH)_3$ Pnam
●**MAGNOLITE** Hg_2TeO_3 Pbm2
NATRITE γ-Na_2CO_3 Mon.
●**PARATACAMITE** $Cu_2Cl(OH)_3$ $R\bar{3}$
PENFIELDITE $Pb_2Cl_3(OH)$ P6/m...
RADHAKRISHNAITE $PbTe_3(Cl,S)_2$ P4/m...
●**RASVUMITE** KFe_2S_3 Cmcm
STERNBERGITE $AgFe_2S_3$ Cmma
SULPHOTSUMOITE Bi_3Te_2S $P\bar{3}m1$
UYTENBOGAARDTITE Ag_3AuS_2 $P4_122$...
ZABUYELITE Li_2CO_3 C2/c

Table 15S

AB_2C_4

CLOSE-PACKED

STRUCTURE TYPE	POPULATION	DERIVATIVES
FORSTERITE $Mg_2{}^{o}Si^{t}[O_4]^{h}$ Pbnm	CHRYSOBERYL $Al_2{}^{o}Be^{t}[O_4]^{h}$ FAYALITE $Fe_2{}^{o}Si^{t}[O_4]^{h}$ LIEBENBERGITE $(Ni,Mg)_2{}^{o}Si^{t}[O_4]^{h}$ TEPHROITE $Mn_2{}^{o}Si^{t}[O_4]^{h}$	
SPINEL $Al_2{}^{o}Mg^{t}[O_4]^{c}$ $Fd\bar{3}m$	CARROLLITE $(Co,Ni)_2{}^{o}Cu^{t}[S_4]^{c}$ CUPROIRIDSITE $Ir_2{}^{o}Cu^{t}[S_4]^{c}$ CUPRORHODSITE $Rh_2{}^{o}Cu^{t}[S_4]^{c}$ DAUBRÉELITE $Cr_2{}^{o}Fe^{t}[S_4]^{c}$ FLETCHERITE $(Ni,Co)_2{}^{o}Cu^{t}[S_4]^{c}$ GREIGITE $Fe_2{}^{o}Fe^{t}[S_4]^{c}$ KALININITE $Cr_2{}^{o}Zn^{t}[S_4]^{c}$ POLYDYMITE $Ni_2{}^{o}Ni^{t}[S_4]^{c}$ INDITE $In_2{}^{o}Fe^{t}\{S_4]^{c}$ SIEGENITE $Ni_2{}^{o}Co^{t}[S_4]^{c}$ BORNHARDTITE $Co_2{}^{o}Co^{t}[Se_4]^{c}$ TRÜSTEDTITE $Ni_2{}^{o}Ni^{t}[Se_4]^{c}$ TYRRELLITE $(Co,Ni)_2{}^{o}(Cu,Ni)^{t}[Se_4]^{c}$ CHROMITE $Cr_2{}^{o}Fe^{t}[O_4]^{c}$ COCHROMITE $(Cr,Al,Fe)_2{}^{o}(Co,Ni,Fe)^{t}[O_4]^{c}$ COULSONITE $V_2{}^{o}Fe^{t}[O_4]^{c}$ CUPROSPINEL $Fe_2{}^{o}(Cu,Mg)^{t}[O_4]^{c}$ FILIPSTADITE $(Mn,Mg)_2{}^{o}(Sb,Fe)^{t}[O_4]^{c}$ FRANKLINITE $(Fe,Mn)_2{}^{o}(Zn,Mn,Fe)^{t}[O_4]^{c}$ GAHNITE $Al_2{}^{o}Zn^{t}[O_4]^{c}$ GALAXITE $Al_2{}^{o}Mn^{t}[O_4]^{c}$ HERCYNITE $Al_2{}^{o}Fe^{t}[O_4]^{c}$ JACOBSITE $(Fe,Mn)_2{}^{o}(Mn,Fe,Mg)^{t}[O_4]^{c}$ LINNAEITE $Co_2{}^{o}Co^{t}[S_4]^{c}$ MAGNESIOCHROMITE $Cr_2{}^{o}Mg^{t}[O_4]^{c}$ MAGNESIOFERRITE $Fe_2{}^{o}Mg^{t}[O_4]^{c}$ MAGNETITE $Fe_2{}^{o}Fe^{t}[O_4]^{c}$ MANGANOCHROMITE $(Cr,V)_2{}^{o}(Mn,Fe)^{t}[O_4]^{c}$ QANDILITE $(Mg,Fe)_2{}^{o}(Ti,Fe,Al)^{t}[O_4]^{c}$ TREVORITE $Fe_2{}^{o}Ni^{t}[O_4]^{c}$ ULVÖSPINEL $Fe_2{}^{o}Ti^{t}[O_4]^{c}$ VIOLARITE $Ni_2{}^{o}Fe^{t}[S_4]^{c}$ VUORELAINENITE $(V,Cr)_2{}^{o}(Mn,Fe)^{t}[O_4]^{c}$ ZINCOCHROMITE $Cr_2{}^{o}Zn^{t}[O_4]^{c}$	DONATHITE $(Cr,Fe)_2{}^{o}(Fe,Mg)^{t}[O_4]^{c}$ $P4_2/nnm$ HETAEROLITE $Mn_2{}^{o}Zn^{t}[O_4]^{c}$ $I\,4_1/amd$ HAUSMANNITE $Mn_2{}^{o}Mn^{t}[O_4]^{c}$ $I\,4_1/amd$ IWAKIITE $(Fe,Mn)_2{}^{o}Mn^{t}[O_4]^{c}$ $I\,4_1/amd$ LAIHUNITE $(Fe_{0.8}{}^{+2},Fe_{0.8}{}^{+3},\square_{0.4})^{o}Si^{t}[O_4]^{c}$ $P2_1/b$ MAGHEMITE $(Fe_{1.67},\square_{0.33})^{o}Fe^{t}[O_4]^{c}$ $P2_13$

CHAIN

MINIUM $Pb_2{}^{[3n]}\{1\infty\}[Pb^{o}O_4]$ $P4_2/mbc$

MINERALS TENTATIVELY CLASSIFIED

ARCANITE $K_2{}^{[9/10]}\{g\}[S^{t}O_4]$ Pnam
GALENOBISMUTITE $Pb^{[7]}Bi_2{}^{[6/7]}[S_4]^{\#}$ Pnam
HEIDEITE $(Ti,Fe)_2{}^{o}(Fe,Cr)_{1+x}{}^{o}[S_4]^{h}$ I 2/m
JOSÉITE-A $Bi_4{}^{o}[TeS_2]^{c/h}$ $R\bar{3}m$
JOSÉITE-B $Bi_4{}^{o}[Te_2S]^{c/h}$ $R\bar{3}m$
PHENAKITE $Be_4{}^{t}Si_2{}^{t}[O_8\square]^{(3R21)\ by}$ $R\bar{3}$
SCHAFARZIKITE $1\infty[Fe^{o}Sb_2{}^{[3n]}O_4]$ $P4_2/mbc$
WILLEMITE $Zn_4{}^{t}Si_2{}^{t}[O_8\square]^{(3R21)\ by}$ $R\bar{3}$

MINERALS NOT YET CLASSIFIED

● **BERTHIERITE** $FeSb_2S_4$ Pnam
BRUNOGEIERITE $Fe_2(Ge,Fe)O_4$ $Fd\bar{3}m$
CHURSINITE Hg_2AsO_4 $P2_1/c$
FLORENSOVITE $Cu(Cr,Sb)_2S_4$ $Fd\bar{3}m$
GUETTARDITE $Pb(Sb,As)_2S_4$ $P2_1/a$
KADYRELITE $Hg_4(Br,Cl)_2O$ $I\,a\bar{3}d$
● **LARNITE** β-Ca_2SiO_4 $P2_1/n$
● **LEITEITE** $ZnAs_2O_4$ $P2_1/c$
● **MAROKITE** $CaMn_2O_4$ Pmab...
MASCAGNITE $(NH_4)_2SO_4$ Pmcn
PLATYNITE $PbBi_2(Se,S)_4$ $R\bar{3}m$
RINGWOODITE $(Mg,Fe)_2SiO_4$ Cub.
● **SARTORITE** $PbAs_2S_4$ $P2_1/n$
● **TARAPACÁITE** K_2CrO_4 Pnam
● **THENARDITE** α-Na_2SO_4 Fddd
● **TRIPPKEITE** $CuAs_2O_4$ $P4_2/mbc$
TWINNITE $Pb(Sb,As)_2S_4$ Pnmn
VOZHMINITE $(Ni,Co)_4(As,Sb)S_2$ Hex.
WADSLEYITE β-$(Mg,Fe)_2SiO_4$ Imma

Table 16S

AB_2C_5

CLOSE-PACKED

KYANITE $Al_2{}^{o}Si^{t}[O_5]^{c}$ $P\bar{1}$

CHAIN

MULLITE $Al_2{}^{o}\{1\infty\}[(Si_{2-2x}{}^{t},Al_{2+x})^{t}O_{10-x}]$ Pbam (disord.deriv. Sillimanite)
PSEUDOBROOKITE $Fe_2{}^{[6]}\{1\infty\}[Ti^{o}O_5]$ C2
SANBORNITE $Ba^{[9]}\{1\infty\}[Si_2{}^{t}O_5]$ Pmcn

MINERALS NOT YET CLASSIFIED

BERDESINSKIITE V_2TiO_5 C2/c...
RAJITE $CuTe_2O_5$ $P2_1/c$
SMIRNITE Bi_2TeO_5 Cm2a
●**STIBIVANITE** Sb_2VO_5 Pmcn
●**STIBIVANITE-2O** Sb_2VO_5 Pmcn

AB_2C_6

CLOSE-PACKED

STRUCTURE TYPE	POPULATION	DERIVATIVES
TAPIOLITE $(Fe,Mn)^{o}(Ta,Nb)_2{}^{o}[O_6]^{h}$ $P4_2/mnm$ (subs. deriv. Rutile)	BYSTRÖMITE $Mg^{o}Sb_2{}^{o}[O_6]^{h}$ FERROTAPIOLITE $Fe^{o}(Ta,Nb)_2{}^{o}[O_6]^{h}$ MANGANOTAPIOLITE $(Mn,Fe)^{o}(Ta,Nb)_2{}^{o}[O_6]^{h}$ ORDOÑEZITE $Zn^{o}Sb_2{}^{o}[O_6]^{h}$ TRIPUHYITE $Fe^{o}Sb_2{}^{o}[O_6]^{h}$	NIOBO-AESCHYNITE-(Ce) $(Ce,Ca)^{o}(Nb,Ti)_2{}^{o}[(O,OH)_6]^{h}$ Pbnm
HIERATITE $Si^{o}[K_2F_6]^{c}$ Fm3m	CRYPTOHALITE $Si^{o}[(NH_4)_2F_6]^{c}$	BARARITE $Si^{o}[(NH_4)_2F_6]^{c}$ C3m MALLADRITE $Si^{o}[Na_2F_6]^{c}$ P321
TANTALITE $(Fe,Mn)^{o}Ta_2{}^{o}[O_6]^{h}$ Pbcn (subs. deriv. Scrutinyite)	EUXENITE-(Y) $(Y,Ca,Ce,U,Th)^{o}(Nb,Ta,Ti)_2{}^{o}[O_6]^{h}$ FERROCOLUMBITE $Fe^{o}Nb_2{}^{o}[O_6]^{h}$ FERSMITE $(Ca,Ce,Na)^{o}(Nb,Ta,Ti)_2{}^{o}[O,OH,F]_6{}^{h}$ MAGNOCOLUMBITE $(Mg,Fe,Mn)^{o}(Nb,Ta)_2{}^{o}[O_6]^{h}$ MANGANOCOLUMBITE $(Mn,Fe)^{o}(Nb,Ta)_2{}^{o}[O_6]^{h}$ MANGANOTANTALITE $(Mn,Fe)^{o}(Ta,Nb)_2{}^{o}[O_6]^{h}$ TANTEUXENITE-(Y) $(Y,Ce,Ca)^{o}(Ta,Nb,Ti)_2{}^{o}[(O,OH)_6]^{h}$	

MINERAIS TENTATIVELY CLASSIFIED

AESCHYNITE-(Ce) $(Ce,Ca,Fe,Th)^{[8]}$ $\{3\infty\}$ $[(Ti,Nb)_2{}^{o}(O,OH)_6]$ Pnma
AESCHYNITE-(Y) $(Y,Ca,Fe,Th)^{[8]}$ $\{3\infty\}$ $[(Ti,Nb)_2{}^{o}(O,OH)_6]$ Pbnm
FOORDITE $Sn^{[8ap]}(Nb,Ta)_2{}^{o}[O_6]^{h}$ C2/c
THOREAULITE $Sn^{[8ap]}Ta_2{}^{o}[O_6]^{h}$ C2/c

MINERALS NOT YET CLASSIFIED

●**BRANNERITE** $(U,Ca,Y,Ce)(Ti,Fe)_2O_6$ C2/m
CHANGBAIITE $PbNb_2O_6$ R3m
●**KOECHLINITE** Bi_2MoO_6 $Pna2_1$
●**LAUTARITE** $Ca(IO_3)_2$ $P2_1/n$
●**LUCASITE-(Ce)** $(Ce,La)Ti_2(O,OH)_6$ I 2/a
LUDLOCKITE $(Fe,Pb)As_2O_6$ P1...
●**NITROBARITE** $Ba(NO_3)_2$ $P2_13$
POLYCRASE-(Y) $Y(Ti,Nb)_2(O,OH)_6$ Pbcn
RUSSELLITE Bi_2WO_6 $Pca2_1$
RYNERSONITE $Ca(Ta,Nb)_2O_6$ Pmnb
TANTAL-AESCHYNITE-(Y) $(Y,Ce)(Ta,Ti,Nb)_2O_6$ Orth.
THORUTITE $(Th,U,Ca)Ti_2(O,OH)_6$ C2/m
VIGEZZITE $(Ca,Ce)(Nb,Ta,Ti)_2O_6$ Pmbn
YTTROTANTALITE-(Y) $(Y,U,Ca)(Ta,Fe)_2(O,OH)_6$ Orth.

Table 17S

AB_2C_8

MINERALS NOT YET CLASSIFIED

FROLOVITE $Ca(B(OH)_4)_2$ $P\bar{1}$
LIANDRATITE $U(Nb,Ta)_2O_8$ $P\bar{3}1m$
PROSOPITE $CaAl_2(F,OH)_8$ C2/c
SEDOVITE $U(MoO_4)_2$ Orth.

Table 18S

$AB_mC_n(m,n\geq3)$

CLOSE-PACKED

STRUCTURE TYPE	POPULATION	DERIVATIVES
ENARGITE $Cu_3{}^tAs^t[S_4]^h$ $Pmn2_1$ (subs.deriv.Wurtzite-2H)		
FAMATINITE $Cu_3{}^tSb^t[S_4]^c$ $I\bar{4}2m$ (subs.deriv.Sphalerite)	LUZONITE $Cu_3{}^tAs^t[S_4]^c$ PERMINGEATITE $Cu_3{}^t(Sb,As)^t[Se_4]^c$	
PENTLANDITE $(Ni,Fe)^o(Ni,Fe)_8{}^t[S_8]^c$ $Fm\bar{3}m$	ARGENTOPENTLANDITE $Ag^o(Fe,Ni)_8{}^t[S_8]^c$ COBALT PENTLANDITE $(Co,Ni,Fe)^o(Co,Ni,Fe)_8{}^t[S_8]^c$ GEFFROYITE $(Cu,Fe,Ag)^o(Cu,Fe,Ag)_8{}^t[(Se,S)_8]^c$ MANGANESE-SHADLUNITE $(Mn,Pb)^o(Fe,Cu)_8{}^t[S_8]^c$ SHADLUNITE $(Pb,Cd)^o(Fe,Cu)_8{}^t[S_8]^c$	KHARAELAKHITE $(Cu,Pt,Pb,Fe,Ni)^o(Cu,Pt,Pb,Fe,Ni)_8{}^t[S_8]^c$ Pmmm..
HOLLANDITE $Mn_8{}^o[(Ba,K)O_{16}]^{chh}$ I 2/m	AKAGANÉITE $Fe_8{}^o[(Cl,H_2O)(O,OH)_{16}]^{chh}$ CORONADITE $Mn_8{}^o[Pb\ O_{16}]^{chh}$ CRYPTOMELANE (mon.) $Mn_8{}^o[K\ O_{16}]^{chh}$	CRYPTOMELANE (tet.) $Mn_8{}^o[K\ O_{16}]^{chh}$ I 4/m PRIDERITE $(Ti,Fe)_8{}^o[(K,Ba)\ O_{16}]^{chh}$ I 4/m

GROUP

PROUSTITE $[[Ag_3]\{g\}[As^{[3n]}S_3]]$ R3c   Pop.: PYRARGYRITE $[[Ag_3]\{g\}[Sb^{[3n]}S_3]]$

FRAMEWORK

SANIDINE $(K,Na)^{[9]}\{3\infty\}[(Si,Al)_4{}^tO_8]$ C2/m

Table 19S

$AB_mC_n(m,n\geq 3)$(cont.)

MINERALS TENTATIVELY CLASSIFIED

ARSENOSULVANITE $Cu_3{}^t(As,V)^t[S_4]^c$ $P\bar{4}3m$ (subs.deriv.Sphal.)
BERNARDITE $Tl^{[8]}\{3\infty\}[As_5S_8]$ $P2_1/c$
BORNITE $(Cu_5Fe\square_2)^t[S_4]^c$ $P42_1c$ (defect deriv.Digenite)
CRICHTONITE $(Ti,Mn)_{19}{}^oFe_2{}^t[(Sr,La,Ce)O_{38}]^{hhc}$ $R\bar{3}$
HIBONITE $(Al,Ti,Mg)_{12}{}^t[(Ca,Ce)O_{19}]^{c/h}$ $P6_3/mmc$
HÖGBOMITE-8H $(Al,Fe,Ti,Zn)_8{}^o\,(Fe,Zn,Mg,Al)_3{}^t[O_{15}\,OH]^{ccch}$ $P6_3m$
HYALOPHANE $(K,Ba)^{[9]}\{3\infty\}[(Al,Si)_4{}^tO_8]$ $C2/m$
LITHIOPHOSPHATE $Li_3{}^tP^t[O_4]^h$ Pmnb
MAGNETOPLUMBITE $(Fe,Mn)_{12}{}^o[PbO_{19}]^{chhhc}$ $P6_3/mmc$
PETRUKITE $(Cu,Fe,Sn)_3{}^tSn^t[S_4]^c$ $Pmn2_1$
SULVANITE $Cu_3{}^tV^t[S_4]^c$ $P\bar{4}3m$

MINERALS NOT YET CLASSIFIED

●**BANNERMANITE** $(Na,K)_xV_6O_{15}$ C2/m
●**BETEKHTINITE** $Cu_{10}(Pb,Fe)S_6$ Immm
BILLINGSLEYITE $Ag_7(Sb,As)S_6$ $C222_1$
BOROVSKITE Pd_3SbTe_4 $Fm\bar{3}m$?
CALCIOTANTITE $CaTa_4O_{11}$ $P6_3$
CAMERONITE $AgCu_7Te_{10}$ $P4_2mmc$
CARLFRIESITE $CaTe_3O_8$ C2/c
CHAMÉANITE $(Cu,Fe)_4As(Se,S)_4$ Cub
CHLORMANGANOKALITE K_4MnCl_6 $R\bar{3}c$
.●**CLIFFORDITE** UTe_3O_9 $Pa\bar{3}$
CROOKESITE Cu_7TlSe_4 I 4
DANIELSITE $(Cu,Ag)_{14}HgS_8$ Orth.
DIAOYUDAOITE $NaAl_{11}O_{17}$ P6/mmc
●**ELLISITE** Tl_3AsS_3 R3m
FURUTOBEITE $(Cu,Ag)_6PbS_4$ Cm...
HATRURITE Ca_3SiO_5 $R\bar{3}m$
LINDSLEYITE $(Ba,Sr)(Ti,Cr,Fe)_{21}O_{38}$ $R\bar{3}$
LITHIOTANTITE $Li(Ta,Nb)_3O_8$ $P2_1$/?
LITHIOWODGINITE $LiTa_3O_8$ C2/c
●**LIVINGSTONITE** $HgSb_4S_8$ A2/a
●**MATHIASITE** $(K,Ca,Sr)(Ti,Cr,Fe,Mg)_{21}O_{38}$ $R\bar{3}$
MGRIITE Cu_3AsSe_3 $Pd\bar{3}m$
MOROZEVICZITE $Pb_3Ge_{1-x}S_4$ Cub.
NAGYÁGITE $Au(Pb,Sb,Fe)_8(Te,S)_{11}$ P4
●**OLYMPITE** Na_3PO_4 Pnma...
OTJISUMEITE $PbGe_4O_9$ P1...
●**PARAPIERROTITE** $Tl(Sb,As)_5S_8$ Pn
●**PAVONITE** $(Ag,Cu)(Bi,Pb)_3S_5$ C2/m
PENZHINITE $(Ag,Cu)_4Au(S,Se)_4$ $P6_322$
PLUMBOFERRITE $PbFe_4O_7$ P312
POLKOVICITE $(Fe,Pb)_3(Ge,Fe)_{1-x}S_4$ Cub.
PYROSTILPNITE Ag_3SbS_3 $P2_1/c$
SABATIERITE Cu_6TlSe_4 Orth.
SANTANAITE $Pb_{11}CrO_{16}$ $P6_322$
SELENOSTEPHANITE $Ag_5Sb(Se,S)_4$ $P2_12_12_1$
●**SILLÉNITE** γ-$Bi_{12}SiO_{20}$ I 23
SKINNERITE Cu_3SbS_3 $P2_1/c$
●**STEPHANITE** Ag_5SbS_4 $Cmc2_1$
STIBICONITE $Sb_3O_6(OH)$ $Fd\bar{3}m$
●**TAKEUCHIITE** $(Mg,Mn)_3BO_5$ Pnnm
TEMAGAMITE Pd_3HgTe_3 Orth.
USTARASITE $Pb(Bi,Sb)_6S_{10}$ (?)
WAKABAYASHILITE $SbAs_{10}S_{18}$ $P2_1/m$
WINSTANLEYITE $TiTe_3O_8$ Ia3
●**WITTICHENITE** Cu_3BiS_3 $P2_12_12_1$
WITTITE $Pb_{0.35}Bi_{0.44}(S,Se)$ P2/m
●**XANTHOCONITE** Ag_3AsS_3 C2/c
●**YAFSOANITE** $(Zn,Ca,Pb)_3TeO_6$ $I\,4_132$
YIMENGITE $K(Cr,Ti,Fe,Mg)_{12}O_{19}$ $P6_3/mmc$
YTTROTUNGSTITE-(Y) (Y,La,Ca) $(W,Fe,Si,Al,Ti)_2\,(O,OH,H_2O)_9$ $P2_1/n$

Table 20S

$A_pB_qC_r(p,q\geq 2;r\geq 3)$

FRAMEWORK

EULYTITE $3\infty[Bi_4^{[3n]}(Si^tO_4)_3]$ $I\bar{4}3d$

MINERALS TENTATIVELY CLASSIFIED

STRUCTURE TYPE	POPULATION
BARIOPYROCHLORE $(Ba,Sr)_2^{cb}(Nb,Ti)_2^{[6]}[(O,OH)_7\square]^{Qs}$ $Fd\bar{3}m$	CALCIOBETAFITE $(Ca,Na)_2^{cb}(Nb,Ti)_2^{[6]}[(O,OH)_7\square]^{Qs}$ KALIPYROCHLORE $(K,H_2O)_2^{cb}Nb_2^{[6]}[(O,OH)_7\square]^{Qs}$ STANNOMICROLITE $(Sn,Fe,Mn)_2^{cb}(Ta,Nb,Sn)_2^{[6]}[(O,OH,F)_7\square]^{Qs}$ STIBIOMICROLITE $(Ca,Sb,Na)_2^{cb}(Ta,Nb)_2^{[6]}[O,OH,F)_7\square]^{Qs}$ BARIOMICROLITE $(Ba\square)^{cb}(Ta,Nb)_2^{[6]}[(O,OH)_7\square]^{Qs}$ BISMUTOSTIBICONITE $(Bi\square)^{cb}(Sb,Fe)_2^{[6]}[O_7\square]^{Qs}$ PLUMBOMICROLITE $((Pb,Na,Ca)\square)^{cb}(Ta,Nb)_2^{[6]}[(O,OH)_7\square]^{Qs}$
BLOSSITE $Cu_2^{[5]}\{g\}[V_2^tO_7]$ Fdd2	
BOULANGERITE $Pb_5^{[6/7]}Sb_4^{[6/7]}[S_{11}]^{\#}$ Pbnm	
CANNIZZARITE $Pb_{46}^{o}Bi_{54}^{o}[S_{127}]^{\#}$ $P2_1/m$	
CHERVETITE $Pb_2^{[8/9]}\{g\}[V_2^tO_7]$ $P2_1/a$	
COSALITE $Pb_2^{o}Bi_2^{[5/6]}[S_5]^{\#}$ Pbnm	
CUPROBISMUTITE $Cu_4^{t/tr}Bi_7^{o}[S_{12}]^{\#}$ C2/m	
DOLORESITE $2\infty[V_3^{o}O_4(OH)_4]$ C2/m	
GODLEVSKITE $(Ni,Fe)_5^{t}(Ni,Fe)_4^{[4n]}[S_8]^c$ C222	
HENRYITE $Cu_4^{t/tr}Ag_3^{o}[Te_4]^c$ Cub.	
KOTOITE $Mg_3^{o}B_2^{tr}[O_6]^h$ Pnmn	JIMBOITE $Mn_3^{o}B_2^{tr}[O_6]^h$
LAMMERITE $Cu_3^{o}(As,P)_2^{t}[O_8]^c$ $P2_1/a$	
LILLIANITE $Pb_3^{o}Bi_2^{o}[S_6]^{\#}$ Bbmm	
LIVEINGITE $Pb_9^{o}[As_{13}S_{28}]^{\#}$ $P2_1$	
LOPEZITE $K_2^{[6/8]}\{g\}[Cr_2^tO_7]$ P1	
MOOIHOEKITE $Cu_9^tFe_9^t[S_{16}]^c$ $P\bar{4}2m$ (deriv. Sphalerite)	
PARAMELACONITE $Cu_2^{sq}Cu_2^{[2]}[O_3]^{Qs}$ $I4_1amd$	
PSEUDORUTILE $Fe_2^{o}Ti_3^{o}[O_9]^h$ $P6_322$	
RANKAMAITE $(Ta,Nb,Al)_{11}^{o}[(Na,K,Pb,Li)_3(O,OH)_{30}]^c$ C222…	
RANKINITE $Ca_3^{[7]}\{g\}[Si_2^tO_7]$ $P2_1/a$	
ROBINSONITE $Pb_4^{[6/8]}Sb_6^{[5/7]}[S_{13}]^{ch}$ P1	
ROMANÈCHITE $Mn_5^{o}[(Ba,H_2O)_2O_{10}]^{cchh}$ C2/m	
SARCOPSIDE $(Fe,Mn,Mg)_3^{o}P_2^{t}[O_8]^h$ $P2_1/a$ (relat.Heterosite)	
SINNERITE $Cu_6^tAs_4^{[3n]}[S_9\square]^c$ P1 (relat. Sphalerite)	
TALNAKHITE $Cu_9^t(Fe,Ni)_8^t[S_{16},\square]^c$ $I\bar{4}3m$ (subs.d.Sphalerite)	
THORTVEITITE $(Sc,Y)_2^{o}\{g\}[Si_2^tO_7]$ C2/m	
YTTRIALITE-(Y) $(Y,Th)_2^{o}\{g\}[Si_2^tO_7]$ $P2_1/m$	
ZIESITE $Cu_2^{[5]}\{g\}[V_2^tO_7]$ C2/c	

Table 21S

$A_pB_qC_r(p,q\geq 2;r\geq 3)$(cont.)

MINERALS NOT YET CLASSIFIED

ASCHAMALMITE $Pb_6Bi_2S_9$ C2/m...
BISMUTOHAUCHECORNITE $Ni_9Bi_2S_8$ P4/mmm
BOGDANOVITE $Au_5(Cu,Fe)_3(Te,Pb)_2$ Cub.
●**BROWNMILLERITE** $Ca_2(Al,Fe)_2O_5$ Ibm2
BURSAITE $Pb_5Bi_4S_{11}$ Bbmm
CHEKHOVICITE $Bi_2Te_4O_{11}$ $P2_1$
●**CHIOLITE** $Na_5Al_3F_{14}$ P4/mmc
CLARKEITE $(Na,Ca,Pb)_2U_2(O,OH)_7$ (?)
COMANCHEITE $Hg_{13}O_9(Cl,Br)_8$ Pnnm...
CORDEROITE α-$Hg_3S_2Cl_2$ $I\,2_13$
●**DIOMIGNITE** $Li_2Bi_4O_7$ $I\,4_1cd$
DONHARRISITE $Ni_9Hg_3S_9$ C2/m...
●**DUFRÉNOYSITE** $Pb_2As_2S_5$ $P2_1$
EDGARBAILEYITE $Hg_6Si_2O_7$ C2/m
ERTIXIITE $Na_2Si_4O_9$ Cub.
●**FARRINGTONITE** $Mg_3(PO_4)_2$ $P2_1/n$
FIEDLERITE $Pb_3Cl_4(OH)_2$ $P2_1/a$
●**FREUDENBERGITE** $Na_2(Ti,Fe)_8O_{16}$ C2/m
●**FÜLÖPPITE** $Pb_3Sb_8S_{15}$ C2/m
●**GEOCRONITE** $Pb_{14}(Sb,As)_6S_{23}$ $P2_1/m$
GORTDRUMITE $(Cu,Fe)_6Hg_2S_5$ Orth.
●**GRATONITE** $Pb_9As_4S_{15}$ R3m
●**HÄGGITE** $V_2O_2(OH)_3$ C2/m
●**HETEROMORPHITE** $Pb_7Sb_8S_{19}$ C2/c
HODRUSHITE $Cu_4Bi_6S_{11}$ A2/m
●**HUTCHINSONITE** $(Tl,Pb)_2As_5S_9$ Pbca
HYDROROMARCHITE $Sn_3O_2(OH)_2$ P4mnc...
●**IMHOFITE** $Tl_6As_{15}S_{25}$ $P2_1/n$
IRTYSHITE $Na_2(Ta,Nb)_4O_{11}$ $P6_3/m$...
JANGGUNITE $(Mn,Fe)_6O_8(OH)_6$ Orth.
JEPPEITE $(K,Ba)_2(Ti,Fe)_6O_{13}$ C2/m
KARIBIBITE $Fe_2As_4(O,OH)_9$ Orth.
KEIVIITE-(Y) $(Y,Yb)_2Si_2O_7$ C2/m
KEIVIITE-(Yb) $(Yb,Y)_2Si_2O_7$ C2/m
KEYITE $(Cu,Zn,Cd)_3(AsO_4)_2$ I 2/m
●**KILCHOANITE** $Ca_3Si_2O_7$ Imam...
KUTÍNAITE $Ag_6Cu_{14}As_7$ $I\,\bar{4}$-3d
KYZYLKUMITE $V_2Ti_3O_9$ Mon.
LAROSITE $(Cu,Ag)_{21}(Pb,Bi)_2S_{13}$ Orth.
LAUNAYITE $Pb_{22}(Sb,As)_{26}S_{61}$ C2...
LAVRENTIEVITE $Hg_3S_2(Cl,Br)_2$ P2/m...
MADOCITE $Pb_{17}(Sb,As)_{16}S_{41}$ Pba2...
MAYENITE $Ca_{12}Al_{14}O_{33}$ I 43d
MCBIRNEYITE $Cu_3(VO_4)_2$ P1
●**MENDIPITE** $Pb_3O_2Cl_2$ Pnma
MILLOSEVICHITE $(Al,Fe)_2(SO_4)_3$ Trig.
NATROSILITE $Na_2Si_2O_5$ $P2_1/a$
●**NATROTANTITE** $Na_2Ta_4O_{11}$ $R\bar{3}c$
●**ONORATOITE** $Sb_8O_{11}Cl_2$ C2/m
●**PARKERITE** $Ni_3(Bi,Pb)_2S_2$ Pnam
●**PAULMOOREITE** $Pb_2As_2O_5$ $P2_1/a$
PEARCEITE $(Ag,Cu)_{16}As_2S_{11}$ C2/m
PIERROTITE $Tl_2(Sb,As)_{10}S_{16}$ $Pna2_1$
PINCHITE $Hg_5O_4Cl_2$ Ibam
PLAYFAIRITE $Pb_{16}(Sb,As)_{18}S_{43}$ $P2_1/m$...
POLYBASITE $(Ag,Cu)_{16}Sb_2S_{11}$ C2/m
●**RATHITE-I** $(Pb,Tl)_3As_5S_{10}$ $P2_1/a$
●**REINERITE** $Zn_3(AsO_3)_2$ Pbam
RHODPLUMSITE $Rh_3Pb_2S_2$ $R\bar{3}m$
SAMARSKITE-(Y) $(Y,Ce,U,Fe)_3(Nb,Ta,Ti)_5O_{16}$ C2/c
●**SCHNEIDERHÖHNITE** $Fe_4As_5S_{13}$ $P\bar{1}$
SCHREYERITE $V_2Ti_3O_9$ Mon.
SEMSEYITE $Pb_9Sb_8S_{21}$ C2/c
●**SHANDITE** $Ni_3Pb_2S_2$ $R\bar{3}m$
SOPCHEITE $Ag_4Pd_3Te_4$ Orth.
SORBYITE $Pb_{19}(Sb,As)_{20}S_{49}$ C2/m
SPIROFFITE $(Mn,Zn)_2Te_3O_8$ Ccm...
SREBRODOLSKITE $Ca_2Fe_2O_5$ Pnma
SUANITE $Mg_2B_2O_5$ $P2_1/a$
TRITOMITE-(Y) $(Y,Ca,Ce)_5(Si,B,Al)_3(O,OH)_{13}$ Hex.
TUCEKITE $Ni_9Sb_2S_8$ P4/mmm
TVALCHRELIDZEITE $Hg_{12}(As,Sb)_8S_{12}$ P1
●**TVEITITE-(Y)** $Ca_{14}Y_5F_{43}$ $R\bar{3}$...
VEENITE $Pb_2(Sb,As)_2S_5$ $P2_1cn$...
XANTHIOSITE $Ni_3(As,O_4)_2$ $P2_1/a$
XILINGOLITE $Pb_{3.3}Bi_{1.8}S_6$ C2/m...
●**ZINKENITE** $Pb_9Sb_{22}S_{42}$ $P6_3$

AMORF MINERALS

LECHATELIERITE SiO_2
YTTROPYROCHLORE-(Y) $(Y,Ce,Nd,Th)(Nb,Ta)_2(O,OH)_7$ (metamict)

Table 22S

$A_pB_qC_rD_s\ldots$

MINERALS WHICH ARE TRANSFERRED TO PART II

AERUGITE $Ni_{8.5}^{o}As^{o}As_{2}^{t}[O_{16}]^{c}$ $R\bar{3}m$
ANDALUSITE $Al^{o}Al^{[5]}Si^{t}[O_{5}\square]^{R21my}$ Pnnm
ANDESINE $(Na,Ca)^{[8]}\{3\infty\}[Si_{2}^{t}(Si,Al)^{t}Al^{t}O_{8}]$ P1
ARGYRODITE $Ag_{3}^{t}Ag_{4}^{tr}Ag^{[2]}Ge^{t}[S_{6}\square_{2}]^{c}$ $Pna2_1$
ARMALCOLITE $(Mg,Fe)^{[6]}Ti^{[6]}\{1\infty\}[Ti^{o}O_{5}]$ Bbmm
AUGITE $Ca^{[8]}(Mg,Fe)^{o}\{1\infty\}[(Si,Al)_{2}^{t}O_{6}]$ C2/c
BEUSITE $(Mn,Fe,Ca,Mg)^{[7]}(Mn,Fe,Ca,Mg)_{2}^{[5]}\{g\}[P^{t}O_{4}]_{2}$ $P2_1/c$
BRAUNITE $Mn^{cb}Mn_{6}^{[6]}Si^{t}[O_{12}\square_{4}]^{Qs}$ $I\,4_1/acd$
BYTOWNITE $(Ca,Na)^{[8]}\{3\infty\}[Si_{2}^{t}Al_{2}^{t}O_{8}]$ P1
CALCIBORITE $Ca^{[8]}\{1\infty\}[B^{tr}B^{t}O_{4}]$ Pnca
CANFIELDITE $Ag_{3}^{t}Ag_{4}^{tr}Ag^{[2]}Sn^{t}[S_{6}\square_{2}]^{c}$ $Pna2_1$
CLINOENSTATITE $Mg^{[6]}Mg^{o}\{1\infty\}[Si_{2}^{t}O_{6}]$ $P2_1/c$
CLINOFERROSILITE $(Fe,Mg)_{2}^{[6]}\{1\infty\}[Si_{2}^{t}O_{6}]$ $P2_1/c$
CRYOLITE $Na^{o}Al^{o}[Na_{2}F_{6}]^{c}$ $P2_1/n$
DAVIDITE-(Y) $(Ti,Fe)_{19}^{o}(Ti,Fe)_{2}^{t}[YO_{38}]^{chh}$ $R\bar{3}$
ENSTATITE $Mg^{[6]}Mg^{o}\{1\infty\}[Si_{2}^{t}O_{6}]$ P1
FERROSILITE $(Fe,Mg)^{[6]}(Fe,Mg)^{o}\{1\infty\}[Si_{2}^{t}O_{6}]$ Pbca
GRAFTONITE $(Fe,Mn,Ca)^{[7]}(Fe,Mn,Ca)_{2}^{[5]}\{g\}[P^{t}O_{4}]_{2}$ $P2_1/c$
HAYCOCKITE $Cu_{4}^{t}Fe_{4}^{t}Fe^{o}[S_{8}]^{c}$ P222...
KAMIOKITE $Fe^{o}Fe^{t}Mo_{3}^{o}[O_{8}]^{ch}$ $P6_3mc$
KURAMITE $Cu_{2}^{t}Cu^{t}Sn^{t}[S_{4}]^{c}$ $I\,\bar{4}2m$
LABRADORITE $(Ca,Na)^{[8]}\{3\infty\}[Si_{2}^{t}(Si,Al)^{t}Al^{t}O_{8}]$ C1
LOVERINGITE $(Ti,Fe,Cr)_{19}^{o}(Ti,Fe,Cr)_{2}^{t}[(Ca,Ce,La)O_{38}]^{chh}$ $R\bar{3}$
NOLANITE $(V,Fe,Ti)_{8}^{o}(V,Fe,Ti)_{2}^{t}[O_{14}(OH)_{2}]^{ch}$ P6mc
OLIGOCLASE $(Na,Ca)^{[8]}\{3\infty\}[Si_{2}^{t}(Si,Al)_{2}^{t}O_{8}]$ C1
PIGEONITE $(Ca,Fe)^{[7]}(Mg,Fe)^{o}\{1\infty\}[Si_{2}^{t}O_{6}]$ $P2_1/c$
PLAGIOCLASE $(Na,Ca)^{[8]}\{3\infty\}[Si_{2}^{t}(Si,Al)^{t}Al^{t}O_{8}]$ P1
PLAGIONITE $Pb_{4}^{o}Pb^{[6]}Sb_{6}^{sq}Sb_{2}^{tr}[S_{17}]^{\#}$ C2/c
SAKURAIITE $Cu_{2}^{t}(Zn,Fe)^{t}(In,Sn)^{t}[S_{4}]^{c}$ $Pm\bar{3}m$
SENAITE $(Ti,Fe,Mn)_{19}^{o}(Ti,Fe,Mn)_{2}^{t}[PbO_{38}]^{hhc}$ R3
SILLIMANITE $Al^{o}\{1\infty\}[Si^{t}Al^{t}O_{5}]$ Pbnm
TETRAHEDRITE $3\infty[(Cu,Fe)_{6}^{t}(Cu,Fe)_{6}^{tr}Sb_{4}^{[3n]}S_{13}]$ $I\,\bar{4}3m$

Pop.: GIRAUDITE $3\infty[(Cu,Zn,Ag)_{6}^{t}(Cu,Zn,Ag)_{6}^{tr}(As,Sb)_{4}^{[3n]}(Se,S)_{13}]$
GOLDFIELDITE $3\infty[Cu_{6}^{t}Cu_{6}^{tr}(Te,Sb,As)_{4}^{[3n]}S_{13}]$
HAKITE $3\infty[(Cu,Hg,Ag)_{6}^{t}(Cu,Hg,Ag)_{6}^{tr}Sb_{4}^{[3n]}(Se,S)_{13}]$
TENNANTITE $3\infty[(Cu,Fe)_{6}^{t}(Cu,Fe)_{6}^{tr}As_{4}^{[3n]}S_{13}]$

Table 23S MINERAL STRUCTURE TYPES corresponding to general formulas A, A_m,B_n, and $A_pB_qC_r$

	CLOSE-PACKED	GROUP
A	**OSMIUM** $[Os]^{h}$ **COPPER** $[Cu]^{c}$ **IRON** $[Fe]^{b}$	**ROSICKYITE** $\{g\}[S_8]^{h}$ **SULPHUR** $\{g\}[S_8]$
AB	**TETRAAURICUPRIDE** $[CuAu]^{c}$ **WURTZITE-2H** $Zn^{t}[S]^{h}$ **HONGSHIITE** $[PtCu]^{c}$ **SPHALERITE** $Zn^{t}[S]^{c}$ **WAIRAUITE** $[CoFe]^{b}$ **LITHARGE** $O^{t}[Pb]^{c}$ **CAROBBIITE** $[KF]^{Qs}$ **SAL AMMONIAC** $(NH_4)^{cb}[Cl]^{Qs}$ **NICKELINE** $Ni^{o}[As]^{h}$ **COOPERITE** $(Pt,Pd,Ni)^{sq}[S]^{Qs}$ **HALITE** $Na^{o}[Cl]^{c}$ **VYSOTSKITE** $(Pd,Ni)^{sq}[S]^{N21\ s}$	**REALGAR** $\{g\}[As_4{}^{[3]}S_4]$
AB_2	**CHALCOCITE** $Cu_2{}^{t/tr}[S]^{h}$ **MELONITE** $Ni^{o}[Te_2]^{h}$ **DIGENITE** $Cu_2{}^{t/tr}[S]^{c}$ **CHLOROMAGNESITE** $Mg^{o}[Cl_2]^{c}$ **RUTILE** $Ti^{o}[O_2]^{h}$ **FLUORITE** $Ca^{cb}[F_2]^{Qs}$ **SCRUTINYITE** $Pb^{o}[O_2]^{h}$ **ANYUIITE** $Au^{acb}[Pb_2]^{N21\ f}$ **ANATASE** $Ti^{o}[O_2]^{c}$ **MOLYBDENITE-2H** $Mo^{p}[S_2]^{(2Ts)h}$ **BROOKITE** $Ti^{o}[O_2]^{ch}$ **MOLYBDENITE-3R** $Mo^{p}[S_2]^{(2Ts)c}$ **RAMSDELLITE** $Mn^{o}[O_2]^{h}$ **MALDONITE** $[[Au_2]\ [Bi]]^{(\sim TK)c}$	**PYRITE** $Fe^{o}\{g\}[S_2]^{c}$ **MARCASITE** $Fe^{o}\{g\}[S_2{}^{«h»}]^{c}$
AB_3	**DYSCRASITE** $[Ag_3Sb]^{h}$ **MOLYBDITE** $Mo^{o}[O_3]^{c}$ **AURICUPRIDE** $[Cu_3Au]^{c}$ **GIBBSITE** $Al^{o}[(OH)_3]^{(2Ts)h}$ **MOLYSITE** $Fe^{o}[Cl_3]^{h}$ **DZHALINDITE** $In^{o}[\square\ (OH)_3]^{c}$	**SKUTTERUDITE** $Co_4{}^{o}\{g\}[As_4]_3$
A_2B_3	**CORUNDUM** $Al_2{}^{o}[O_3]^{h}$ **BIXBYITE** $Mn_2{}^{[6]}[O_3\square]^{Qs}$ **TELLUROBISMUTHITE** $Bi_2{}^{o}[Te_3]^{chh}$ **STIBNITE** $Sb_2{}^{[3+4]}[S_3]^{\#}$	
AB_4	**ROALDITE** $N^{o}[(Fe,Ni)_4]^{c}$	
ABC	**MANGANITE** $Mn^{o}[O(OH)]^{h}$ **DIASPORE** $Al^{o}[O(OH)]^{h}$ **HETEROGENITE-2H** $Co^{o}[O(OH)]^{(2Ts)h}$ **HETEROGENITE-3R** $Co^{o}[O(OH)]^{(2Ts)c}$ **MATLOCKITE** $Pb^{[9]}[[F][Cl]]^{(Q2,2Q1')f}$	**ULLMANNITE** $Ni^{o}\{g\}[SbS]^{c}$ **COBALTITE** $Co^{o}\{g\}[AsS]^{o}$ **ARSENOPYRITE** $Fe^{o}\{g\}[AsS^{«h»}]^{c}$
ABC_2	**CHALCOPYRITE** $Cu^{t}Fe^{t}[S_2]^{c}$ **MATILDITE** $Ag^{o}Bi^{o}[S_2]^{c}$	
ABC_3	**ILMENITE** $Fe^{o}Ti^{o}[O_3]^{h}$ **PEROVSKITE** $Ti^{o}[CaO_3]^{c}$	**CALCITE** $Ca^{sp}\{g\}[C^{tr}O_3{}^{«h»}]^{c}$ **ARAGONITE** $Ca^{sp}\{g\}[C^{tr}O_3]$ **VATERITE** $Ca^{[8]}\{g\}[C^{tr}O_3]^{h}$
ABC_4	**WOLFRAMITE** $(Fe,Mn)^{o}W^{o}[O_4]^{h}$ **ZIRCON** $Zr^{do}Si^{t}[O_4]^{*}$ **HETEROSITE** $(Fe,Mn)^{o}P^{t}[O_4]^{h}$ **SCHEELITE** $Ca^{do}W^{t}[O_4]^{*}$ **ANHYDRITE** $Ca^{[8]}S^{t}[O_4]^{*}$ **MONAZITE-(Ce)** $(Ce,La,Nd,Th)^{[9]}P^{t}[O_4]^{*}$	**BARITE** $Ba^{[12]}\{g\}[S^{t}O_4]$
AB_2C_2	**TETRADYMITE** $Bi_2{}^{o}[Te_2S]^{chh}$	
AB_2C_3	**CUBANITE** $Cu^{t}Fe_2{}^{t}[S_3]^{h}$ **COVELLITE** $Cu_2{}^{t}Cu^{tr}[S_3]^{(3Th)s}$	
AB_2C_4	**FORSTERITE** $Mg_2{}^{o}Si^{t}[O_4]^{h}$ **SPINEL** $Al_2{}^{o}Mg^{t}[O_4]^{c}$	
AB_2C_5	**KYANITE** $Al_2{}^{o}Si^{t}[O_5]^{c}$	
AB_2C_6	**TAPIOLITE** $(Fe,Mn)^{o}(Ta,Nb)_2{}^{o}[O_6]^{h}$ **HIERATITE** $Si^{o}[K_2F_6]^{c}$ **TANTALITE** $(Fe,Mn)^{o}Ta_2{}^{o}[O_6]^{h}$	
AB_3C_3		**PROUSTITE** $[[Ag_3]\{g\}[As^{[3n]}S_3]]$
AB_3C_4	**ENARGITE** $Cu_3{}^{t}As^{t}[S_4]^{h}$ **FAMATINITE** $Cu_3{}^{t}Sb^{t}[S_4]^{c}$	
AB_4C_8		
AB_8C_8	**PENTLANDITE** $(Ni,Fe)^{o}(Ni,Fe)_8{}^{t}[S_8]^{c}$	
AB_8C_{16}	**HOLLANDITE** $Mn_8{}^{o}[(Ba,K)\ O_{16}]^{chh}$	
$A_pB_qC_r$		

CHAIN	SHEET	FRAMEWORK
SELENIUM $1\infty[Se^{[2]}]$	**GRAPHITE-2H** $2\infty[C^{tr}]^{Hh}$ **GRAPHITE-3R** $2\infty[C^{tr}]^{Hc}$ **ARSENIC** $2\infty[As^{[3n]}]^{\sim Hc}$ **ARSENOLAMPRITE** $2\infty[As]$	**LONSDALEITE** $3\infty[C^{t}]^{\sim Hh}$ **DIAMOND** $3\infty[C^{t}]^{\sim Hc}$
MONTROYDITE $1\infty[Hg^{[2]}O]$ **CINNABAR** $1\infty[Hg^{o}S^{\text{«Tv»}}]$ **CALOMEL** $1\infty[Hg^{[5]}Cl^{\text{«(2Qe)f»}}]$		**MILLERITE** $3\infty[Ni^{[5]}S]$
		CUPRITE $3\infty[Cu_2^{[2]}O^{t}]$ **TRIDYMITE** $3\infty[Si^{t}O_2^{[2]\text{«ch»}}]^{h}$ **CRISTOBALITE** $3\infty[Si^{t}O_2^{[2]\text{«c»}}]^{c}$ **QUARTZ** $3\infty[Si^{t}O_2^{\text{«Tv»}}]$ **COESITE** $3\infty[Si^{t}O_2]^{c}$ **ICE** $3\infty[O^{t}H_2^{[2]\text{«ch»}}]^{h}$
VALENTINITE $1\infty[Sb_2O_3]$	**ORPIMENT** $2\infty[As_2^{[3n]}S_3]$	
PATRÓNITE $1\infty[VS_4]$		
	DELAFOSSITE $Fe^{o}\{2\infty\}[Cu^{[2]}O_2]^{(3Te)\,c}$	
WOLLASTONITE-1T-2M-3T-4T-5T-7T $Ca^{o}\{1\infty\}[Si^{t}O_3]$ **PYROXFERROITE** $(Fe,Mn,Ca)_7^{o}\{1\infty\}[Si_7^{t}O_{21}]$ **PYROXMANGITE** $Mn_7^{o}\{1\infty\}[Si_7^{t}O_{21}]$ **RHODONITE** $(Mn,Fe,Mg,Ca)_5^{[7]}\{1\infty\}[Si_5^{t}O_{15}]$	**GETCHELLITE** $2\infty[Sb^{[3n]}As^{[3n]}S_3^{[2]}]$	
PUCHERITE $Bi^{[8]}\{1\infty\}[V^{o}O_4]$		**BERLINITE** $3\infty\,[Al^{t}P^{t}O_4]$
MULLITE $Al_2^{o}\{1\infty\}[(Si_{2-2x},Al_{2+x})^{t}\,O_{10-x}]$ **PSEUDOBROOKITE** $Fe_2^{[6]}\{1\infty\}[Ti^{o}O_5]$ **SANBORNITE** $Ba^{[9]}\{1\infty\}[Si_2^{t}O_5]$		
		SANIDINE $(K,Na)^{[9]}\{3\infty\}\,[(Si,Al)_4^{t}\,O_8]$
		EULYTITE $3\infty\,[Bi_4^{[3n]}\,(Si^{t}O_4)_3]$

Table 24S

ApBqCr

	CLOSE-PACKED	GROUP	CHAIN	SHEET	FRAMEWORK	TENT.CLASS.	NOT YET CLASS.	TOTAL
A	3 (+23)	2	1 (+1)	4 (+4)	2 (+1)	9	1	51
AB	12 (+70)	2	3 (+1)	----	1 (+1)	8	7	105
AB_2	14 (+39)	2 (+31)	1	1	6(+3)	11	20 (+●2)	130
AB_3	6 (+4)	1	----	----	----	13	8 (+●1)	33
A_2B_3	4 (+7)	----	1	1 (+1)	----	3	7 (+●4)	28
AB_4	1	----	1	----	----	----	----	2
A_3B_4	----	----	----	----	----	5	4	9
A_2B_5	----	----	----	----	----	1	5	6
AB_6	----	----	----	----	----	----	1	1
A_mB_n	----	----	----	----	----	1	19 (+●2)	22
ABC	5(+9)	3(+10)	----	----	----	10	4	41
ABC_2	2(+3)	----	----		----	10	9(+●3)	27
ABC_3	2(+10)	3(+12)	9	1	----	4	6(+●1)	48
ABC_4	6(+29)	1(+4)	1	----	1	6	5	53
ABC_5	----	----	----	----	----	----	1	1
ABC_6	----	----	----	----	----	1(+2)	7	10
AB_2C_2	1(+1)	----	----	----	----	1	8(+●1)	12
AB_2C_3	2(+2)	----	----	----	----	7	10(+●3)	24
AB_2C_4	2(+44)	----	1	----	----	8	11(+●8)	74
AB_2C_5	1	----	3	----	----	----	3(+●2)	9
AB_2C_6	3(+16)	----	----	----	----	4	9(+●5)	37
AB_2C_8	----	----	----	----	----		4	4
AB_mC_n	4(+13)	1(+1)	----	----	1	11	35(+●15)	81
$A_pB_qC_r$	----		----	----	1	29(+6)	51(+●25)	114
Total	68(+270)	15(+58)	21(+2)	7(+5)	12(+5)	142(+10)	235(+●72)	922

Amorphous 2

$A_pB_qC_rD_s$...(Part II) 36

922+2+36=960

x(+y) x(structure types) +y(population)
z(+●w) z(minerals) +●w(minerals with determined structure)

Table 25S

MINERAL STRUCTURE TYPES

	CLOSE-PACKED	GROUP	CHAIN	SHEET	FRAMEWORK	Total	% CLOSE-PACKED
A	3	2	1	4	2	12	25.0%
AmBn	37	4	5	1	7	54	63.5%
ApBqCr	28	8	12	1	3	52	53.8%
Total	68	14	18	6	12	118	58.4%

CLOSE-PACKED MINERALS / MINERALS

	CLOSE-PACKED STRUCTURE TYPES	CLOSE-PACKED TENT. CLASSIFIED	TOTAL CLOSE-PACKED	TOTAL MINERALS	% CLOSE-PACKED
A	26	9	35	51	68.6%
A_mB_n	158	37	195	336	58.0%
$A_pB_qC_r$	146	78	224	535	41.8%
Total	330	125	455	922	49.3%

CRYSTALLOGRAPHIC PARAMETERS OF MINERAL STRUCTURES

	≤15Å	>15Å	Total	% Minerals ≤15Å
A	50	1	51	98.0%
AmBn	290	46	336	86.3%
ApBqCr	394	141	535	73.6%
Total	734	188	922	79.6%

Table 26S

CLOSE-PACKED A , A_mB_n , and $A_pB_qC_r$ MINERAL STRUCTURE TYPES

Substitution				Insertion						
$[XY_3]$	$[XY_2]$	$[XY]$	$[X]$	A[X]		$A_2[X]$	$A[X_2]$	$A[X_3]$	$A_2[X_3]$	$A[X_4]$
$[Ag_3Sb]^h$ DYSCRASITE			$[Os]^h$ OSMIUM	$Ni^o[As]^h$ NICKELINE	$Zn^t[S]^h$ WURTZITE-2H	$Cu_2^{tri}[S]^h$ CHALCO-CITE	$Ti^o[O_2]^h$ RUTILE; $Pb^o[O_2]^h$ SCRUTINYITE; $Mn^o[O_2]^h$ RAMSDELLITE; $Ni^o[Te_2]^h$ MELONITE	$Fe^o[Cl_3]^h$ MOLYSITE	$Al_2^o[O_3]^h$ CORUNDUM	
$[Cu_3Au]^c$ AURI-CUPRIDE		$[CuAu]^c$ TETRAAURI-CUPRIDE; $[Pt\,Cu]^c$ HONGSHIITE	$[Cu]^c$ COPPER	$Na^o[Cl]^c$ HALITE	$Zn^t[S]^c$ SPHALERITE; $O^t[Pb]^c$ LITHARGE	$Cu_2^{tr}[S]^c$ DIGENITE	$Ti^o[O_2]^c$ ANATASE; $Mg^o[Cl_2]^c$ CHLORO-MAGNESITE	$Mo^o[O_3]^c$ MOLYBDITE		$N^o[(Fe,Ni)_4]^c$ ROALDITE
							$Ti^o[O_2]^{ch}$ BROOKITE		$Bi_2^o[Te_3]^{chh}$ TELLURO-BISMUTHITE	
							$Mo^p[S_2]^{(2Te)h}$ MOLYBDENITE-2H; $Mo^p[S_2]^{(2Te)c}$ MOLYBDENITE-3R	$Al^o[(OH)_3]^{(2Te)h}$ GIBBSITE		
		$[CoFe]^b$ WAIRAUITE	$[Fe]^b$ IRON							
		$[KF]^{Cs}$ CAROBBIITE		$(NH_4)^{cb}[Cl]^{Cs}$ SAL AMMONIAC	$(Pt,Pd,Ni)^{sq}[S]^{Cs}$ COOPERITE		$Ca^{cb}[F_2]^{Cs}$ FLUORITE		$Mn_2^{[6]}[O_3\square]^{Cs}$ BIXBYITE	
	$[[Au_2]\,[Bi]]^{(-TK)c}$ MALDONITE				$(Pd,Ni)^{sq}[S]^{N2Ts}$ VYSOTSKITE		$Au^{acb}[Pb_2]^{N2TT}$ ANYUIITE		$Sb_2^{[3+4]}[S_3]^{s}$ STIBNITE	
Insertion and substitution										
				$Ag^oBi^o[S_2]^c$ MATILDITE	$Cu^tFe^t[S_2]^c$ CHALCOPYRITE; $Cu^tFe_2^t[S_3]^h$ CUBANITE; $Cu_3^tAs^t[S_4]^h$ ENARGITE; $Cu_3^tSb^t[S_4]^c$ FAMATINITE		$Mn^o[O(OH)]^h$ MANGANITE; $(Fe,Mn)^oW^o[O_4]^h$ WOLFRAMITE; $Al^o[O(OH)]^h$ DIASPORE; $(Fe,Mn)^o(Ta,Nb)_2{}^o[O_6]^h$ TAPIOLITE; $(Fe,Mn)^oTa_2{}^o[O_6]^h$ TANTALITE	$In^o[\square(OH)_3]^c$ DZHALINDITE	$Fe^oTi^o[O_3]^h$ ILMENITE	$Ti^o[CaO_3]^c$ PEROVSKITE
							$Co^o[O(OH)]^{(2Te)h}$ HETEROGENITE-2H; $Co^o[O(OH)]^{(2Te)c}$ HETEROGENITE-3R		$Bi_2^o[Te_2S]^{chh}$ TETRADYMITE	
Other kinds of insertion and substitution										
$Mg_2{}^oSi^t[O_4]^h$ FORSTERITE; $Al_2{}^oMg^t[O_4]^c$ SPINEL	$(Fe,Mn)^oP^t[O_4]^h$ HETEROSITE; $Al_2{}^oSi^t[O_5]^c$ KYANITE	$(Ni,Fe)^o(Ni,Fe)_8{}^t[S_8]^c$ PENTLANDITE; $Mn_8{}^o[(Ba,K)O_{16}]^{chh}$ HOLLANDITE		$Si^o[K_2F_6]^c$ HIERATITE; $Cu_2{}^tCu^{tr}[S_3]^{(3Th)s}$ COVELLITE		$Zr^{do}Si^t[O_4]$ * ZIRCON; $Ca^{do}W^t[O_4]$ * SCHEELITE	$Ca^{[8]}S^t[O_4]$ * ANHYDRITE; $(Ce,La,Nd,Th)^{[9]}P^t[O_4]$ * $Pb^{[9]}$ MONAZITE-(Ce)	$Pb^{[9]}$ [[F] [Cl]] $^{(Q2,2Q1)T}$ MATLOCKITE		

CHAPTER 3

Conclusions

One of the first conclusions to be drawn from this study is that *close-packed minerals predominate*. This has already been referred to (Lima-de-Faria and Figueiredo, 1976), and confirms the stability principle of Laves that expresses the tendency for the densest packing (Table 25S). This conclusion is also in agreement with Belov (1947) who said: "In spite of the variety of the mineral crystalline world the whole 'mineralogical game' just reduces to various modes of filling gaps in uniform close packings with the various corresponding patterns". Minerals tend to form close packings, and only when special directional bonds (normally covalent) exist in the structure, do they tend to form less dense arrangements, open structures, giving rise to large voids within the packing of the structural units.

The layer description of close-packed mineral structures, as expressed by condensed models, *confirms their simplicity* (Lima-de-Faria, 1994), which is also in agreement with the Laves stability principles, and obeys simple rules (Lima-de-Faria, 1978). It suggests a possible systematic derivation of mineral structures.

The crystallographic parameters of the unit cells of mineral structures are in general small. As has already been noted (Lima-de-Faria, 1994), the unit cell parameters of the simple minerals show in general small values, less than 15 Å (Table 25S). This fact possibly corresponds to a short-range mechanism of mineral crystallization. As a consequence, any plane direction of high density of atoms within the unit cell will repeat at a short distance, explaining why only a small number of layers is necessary for the layer description.

There is a kind of general rule of small numbers observed in the branches of science such as chemistry and crystallography. In chemistry it is well known that compounds are formed by few elements in the proportion of small integral numbers. In crystallography Hauy's law states that the crystallographic indices of the faces of a crystal correspond to small integral numbers. It seems that variety in nature is curtailed by rules which limit it. *Many structures are derivatives of a certain basic structure*, and may readily be derived from it. Thus, a reduction will be effected on the number of structure types to be memorized.

References

Angelova, O. and Petrova, R. (1996) (Private communication).

Barlow, W. (1898) "XXVI. Geometrische Untersuchung über eine mechanische Ursache der Homogeneität der Structur und Symmetrie: mit besonderer Anwendung auf Krystallisation und chemische Verbindung". *Zeit. Krist. Mineral.*, *29*, 433–588 (Fig. 8, p. 453).

Belov, N.V. (1947) *Structures of ionic crystals and metallic phases* (in Russian) Izd. Akad. Nauk SSSR, Moscow. Quoted by I. Kostov (1981) "Structural patterns and structural genetic trends in minerals". *Soviet Phys. Cryst.*, *26*, 707–708.

Bergerhoff, G. (1995) "Diamond" computer program.

Bergerhoff, G. (1997) ICSD (Inorganic Crystal Structures Database) computer program.

Bergerhoff, G., Bernbt, M., Brandenbury, K. and Degen, T. (1999) "Concerning inorganic crystal structure types". *Acta Cryst.*, *B55*, 147–156.

Blackburn, W.H. and Dennen, W.H. (1997) *Encyclopedia of mineral names*. The Canadian Mineralogist Special Publication 1.

Bokii, G.B. (1954) *Introduction to crystal chemistry* (in Russian). Izdatel'stvo Moskovstogo Universiteta, Moscow. English translation United States Joint Publ. Research Service, New York, 1960.

Bragg, W.L. (1913) "The structure of some crystals as indicated by their diffraction of X-rays". *Proc. R. Soc. London Ser. A*, *89*, 248–277.

Bragg, W.L. (1929) "Atomic arrangement in the silicates". In *Crystal structure and chemical constitution. A general Faraday discussion held by the Faraday Society* (pp. 291–314).

Bragg, W.L. (1930) "The structure of silicates". *Zeit. Krist.*, *74*, 237–305.

Buerger, M.J. (1947) Derivative crystal structures. *J. Chem. Physics*, *15*, 1–16.

Burns, P.C. and Hawthorne, F.C. (1993) "Tolbachite, $CuCl_2$, the first example of Cu^{2+} octahedrally coordinated by Cl^-". *Amer. Miner.*, *78*, 187–189.

Dana, James Dwight (1850) *The system of mineralogy. Third edition*. John Wiley, New York.

Dana, James Dwight (1854) *The system of mineralogy. Fourth edition*. John Wiley, New York.

De Jong, W.F. (1959) *General crystallography. A brief compendium*. Freeman, San Francisco.

Fedorov, E.S. (1913) "Mineralogy". *Encyclopedia Assoc. Granat.*, *28*, 681. Quoted by Povarennyk (1972) p. 15.

Figueiredo, M.O. (1976) (Private comunication on condensed models).

Garrido, J. and Orland, J. (1946) *Los rayos-X y la estructura fina de los cristales*. Dossat, Madrid.

Gottardi, G. (1984) "The classification of minerals". *Acta Cryst.*, *A 40*, Supplement, C-208 (Abstracts ECM-13, Hamburg).

Hawthorne, F.C. (1983) "Graphical enumeration of polyhedral clusters". *Acta Cryst.*, *A39*, 724–736.

Hawthorne, F.C. (1984) "Towards a structural classification of minerals". Presented at the *XIII International Congress of Crystallography*, Hamburg (Abstract). *Acta Cryst.*, *A40*, Supplement, C-245.

Hawthorne, F.C. (1985) "Towards a structural classification of minerals. The $^{VI}M\ ^{IV}T_2\ \Phi_n$ minerals". *Amer. Miner.*, *70*, 455–473.

Hellner, E. (1958) "A structural scheme for sulphide minerals". *Journ. Geol.*, *66*, 503–525.

Hellner, E. (1979) "The frameworks (Bauverbande) of the cubic structure types". In *Structure and Bonding, vol. 37*, 61–140. Springer Verlag, Berlin.

Hölzel. A., Editor (1989) *Systematics of minerals.* Mainz.

Hölzel, A. (1996) *Supplement to the systematic of minerals*.

Kitaigorodskii, A.I. (1955) *Organic chemical crystallography* (in Russian). Press of the Academy of Sciences of the URSS, Moscow. English translation and revision 1961, Consultants Bureau Enterprises, New York.

Kostov, I. (1968) *Mineralogy*. Oliver and Boyd, Edinburgh.

Kostov, I., and Breskovka, V. (1989) *Phosphate, arsenate, and vanadate Minerals. Crystal chemistry and classification*. Kliment Ohridski University Press, Sofia.

Krebs, H. (1968) *Fundamentals of inorganic crystal chemistry*. McGraw-Hill, London.

Laves, F. (1930) "XVI. Die Bau-Zusammenhänge innerhalb der Kristalstrukturen". *I Teil. Zeit. Krist. 73,* 202–265.

Laves, F. (1956) "Crystal structure and atomic size". In *Theory of alloy phases*. American Soc. of Metals, Cleveland, USA.

Liebau. F. (1962) "Die Systematic der Silikate". *Naturwissenschaften, 21,* 481–491.

Liebau, F. (1985) *Structural chemistry of silicates*. Springer-Verlag. Berlin.

Lima-de-Faria, J. (1965) "Systematic derivation of inorganic close-packed structures: AX and AX_2 compounds, sequence of equal layers". *Zeit. Krist., 122,* 259–374.

Lima-de-Faria, J. (1978) "Rules governing the layer organization of inorganic crystal structures". *Zeit. Krist., 148,* 1–5.

Lima-de-Faria, J. (1983) "A proposal for a structural classification of minerals". *Garcia de Orta, Sér. Geologia, 6,* 1–14.

Lima-de-Faria, J. (1988) "The hierarchy of symmetry". Presented at the *XI European Crystallographic Meeting (Vienna).* Abstract in *Zeit. Krist. 185,* 286.

Lima-de-Faria, J. (1991) "On the problem of the measure of the symmetry of crystal structures". *Garcia de Orta, Sér. Geologia, 14,* 45–50.

Lima-de-Faria, J. (1994) *Structural mineralogy. An introduction*. Kluwer Academic Publishers, Dordrecht.

Lima-de-Faria, J. (1998) "Past, present and future of the classification of minerals". *Proceedings of the Symposium on the History of Mineralogy, Petrology and Geochemistry, Munich 1996*. Institut für Geshichte der Naturwissenschaften, München, 17–28.

Lima-de-Faria, J. and Figueiredo, M.O. (1969) "Systematic derivation of inorganic structure types: X_mY_n and A_mX_n, X and Y in cubic and hexagonal close packings, A in octahedral voids". *Zeit. Krist., 130,* 54–67.

Lima-de-Faria, J. and Figueiredo, M.O. (1976) "Classification, notation and ordering on a table of inorganic structure types". *J. Solid State Chem., 16,* 7–20.

Lima-de-Faria, J. and Figueiredo, M.O. (1978) "General chart of stuctural units and building units". *Garcia de Orta, Sér. Geologia, 2,* 69–76.

Lima-de-Faria, J., Hellner, E., Liebau, F., Makovicky, E. and Parthé, E. (1990) "Nomenclature of inorganic structure types. Report of the International Union of Crystallography Commission on Crystallographic Nomenclature, Sub-Committee on the Nomenclature of inorganic structure types". *Acta Cryst., A46,* 1–11.

Linnaeus, C. von (1735) *Systema naturae, sive regna tria naturae systematice proposita per classes, ordinis, genera et species*. Theodorum Haak, Leiden.

Lonsdale, K. (1967) "Origins of the science of crystals". *Chemistry and Industry*, January, 57–58.

Machatschki, F. (1928) "Zur Frage der Struktur und Konstitution der Feldspate". *Cbl. Geol. Paläont. Abteilung A. Miner. Petr.*, 97–104.

Machatschki, F. (1947) "Konstitutionsformeln für den festen Zustand". *Monatsch. Chem., 77,* 333–342.

Machatschki, F. (1953) *Spezielle Mineralogie auf geochemischer Grundlage mit einem Anhang ein kristallchemisches Mineralsystem*. Springer–Verlag, Wien.

Makovicky, E. (1985) "The building principles and classification of sulphosalts based on the SnS archetype". *Fortschr. Miner., 63,* 45–89.

Makovicky, E. (1997a) "Modular crystal chemistry of sulphosalts and other complex sulphides". *EMU notes in mineralogy, 1,* chapter 3, 237–271.

Makovicky, E. (1997b) "Modularity – different types and approaches". *EMU notes in mineralogy, 1,* chapter 5, 315–343.

Megaw, H. (1973) *Crystal structures. A working approach*. W.B. Saunders Company, Philadelphia.

Moore, P. (1992) "Betapakdalite unmasked, and a comment on bond valences". *Aust. J. Chem., 45,* 1335–1354.

Moore, P. (1995) "Closest-packed Mineral Structures of Franklin-Ogdensburg: Kepler's Gift on the Snowflake" (private communication).

Náray-Szabó, St. v. (1930) "Ein auf der Kristallstruktur basierendes Silicasystem". *Zeit. Phys. Abteilung B, 2,* 356–377.

Nickel, E.H. and Nichols, M.C. (1991) *Mineral reference manual*. Chapman & Hall, New York.

Niggli, P. (1945) *Grundlagen der Stereochemie*. Verlag Birkhäuser, Basel. French translation "*Les bases de la Stéreochemie*", Dunod, Paris, 1952.

O'Keeffe, M. and Hyde, M.C. (1996) Crystal structures I. Patterns and symmetry. Mineralogical Society of America, Washington.

Orcel, J. (1954) "Essai sur le concept d´éspèce et les classifications en minéralogie et pétrographie". In *Minéralogie, Cristallographie, Aspects actuels, Soc. Miner. Crist. de France*, 397–432.

Pearson, W.B. (1967) *Handbook of lattice spacings and structures of metals*. Pergamon Press, New York.

Pearson, W.B. (1972) *The crystal chemistry and physics of metals and alloys*. Wiley-Interscience, New York.

Povarennykh, A.S. (1966) *Crystal chemical classification of minerals*. (in Russian) Naukova Dunka, Kiev. English translation, Plenum Press, New York, 1972.

Rentzeperis, P.J. (1963) "The crystal structure of hodgkinsonite Zn_2Mn $[(OH)_2$ $SiO_4]$". *Zeit. Krist., 119,* 117–138.

Roberts, W.L., Rapp, Jr, G.R. and Weber, J. (1974) *Encyclopedia of minerals*. Van Nostrand, New York.

Ross, V. (1958, 1968) "Geochemistry, crystal structure and mineralogy of sulphides". *Economic Geology*, *52*, 755–774.

Spronsen, J.W. van (1969) *The periodic system of the chemical elements. An history of the first hundred years*. Elsevier, New York.

Strunz, H. (1941) *Mineralogische Tabellen. Eine Klassifizierung der Mineralien auf kristallchemischen Grundlage. Mit einer Einfürung in die Kristallchemie*. Akademische Verlagsgesellschaft, Leipzig.

Strunz, H. (1982) *Mineralogical Tabellen*. Akademische Verlagsgesellschaft, Leipzig

Wells, A.F. (1962) *Structural inorganic chemistry*. Clarendon Press, Oxford.

Wells, A.F. (1984) *Structural inorganic chemistry. Fifth edition*. Clarendon Press, Oxford.

Whewell, W. (1837) *History of the inductive sciences. First edition*, London, *3*, 227.

Wyckoff, R.W.G. (1963–1968) *Crystal structures*. John Wiley Interscience Publishers, vol. 1, 1963; vol. 2, 1964; vol. 3, 1965; vol. 4, 1968.

Zoltai, T. (1960) "Classification of silicates and other minerals with tetrahedral structures". *Amer. Min.*, *45*, 960–973.

Abbreviations used in the tables

Abs. = Abstract
Acta Cryst. = *Acta Crystallographica*
Acta Min. Sin. = *Acta Mineralogica Sinica*
Am. Min. = *American Mineralogist*
Angelova & Petrova = Olyana Angelova and Rosica Petrova, 1996 (private communication).
Bokii = *Introduction to crystal chemistry*, by G.B. Bokii, 1960
Bull. Min. = *Bulletin de la Societé française de Mineralogie et Cristallographie*
Can. Min. = *Canadian Mineralogist*
CIM SV23=CIM Special volume 23, "Platinum Group Elements: Mineralogy, Geology, Ricovery", Cabri editor.
Dist. deriv. = Distortion derivative
Dokl. Akad. Sci. USSR = Doklady Akademy of Sciences USSR
Dokl. Earth Sci. = Transations (Doklady) of the USSR Academy of Sciences, Earth Science Section
Subs. deriv. = Substitution derivative
Econ.Geol. = *Economic Geology*
Encyc. Miner. Nam. = *Encyclopedia of mineral names*, by W.H. Blackburn and W.H. Dennen,1997. *The Canadian Mineralogist*, Special Publication 1.
Euro. J. Min. = *European Journal of Mineralogy*
G.O., Sér. Geol. = *Garcia de Orta, Série de Geologia*, Instituto de Investigação Científica Tropical, Lisboa
Hellner,1979 = E. Hellner (1979) "The frameworks (Bauverbände) of the cubic structure types". In *Structure and Bondiing*, *37*, 61–140. Springer Verlag, Berlin.
Hölzel = A.R. Hölzel, Editor (1989) *Systematics of minerals*, Mainz.
Hölzel, suppl. = Hölzel, *Supplement to the systematics of minerals*, 1996.
ICSD = *Inorganic crystal structures database*, Bergerhoff, G., 1997
J. Appl. Cryst. = *Journal of Applied Crystallography*
JCPDS = Joint Committee on Powder Diffraction Standards – International Centre for Diffraction Data
J. Sol. St. Chem. = *Journal of Solid State Chemistry*
Kostov = I. Kostov (1968) *Mineralogy*. Oliver and Boyd, Edinburgh.
K/B = I. Kostov and V. Breskovska (1989) *Phosphate, arsenate and vanadate minerals. Crystal chemistry and classification*. Kliment Ohridski University Press, Sofia.
K/S = I. Kostov and J.M. Stefanova (1982) *Sulphide minerals. Crystal chemistry, parageneses and sytematics*. E. Schweizerbart'sche Verlagsbuchhandlung, Stuttgart.
Krebs = H. Krebs (1968) *Fundamentals of inorganic crystal chemistry*. English translation, McGraw Hill, London.
LF = J. Lima-de-Faria (1994) *Structural mineralogy. An introduction*. Kluwer Akademic Publishers, Dordrecht.
Liebau = F. Liebau (1985) *Structural chemistry of silicates*. Springer Verlag, Berlin.
Megaw = H. Megaw (1973) *Crystal structures. A working approach*. W.B. Saunders Company, London.
Min. Abs. = *Mineralogical Abstracts*
Min. Mag. = *Mineralogical Magazine*
Min. Rec. = *Mineralogical Record*
Min. Zhurn. = *Mineralogical Zhurnal*
Mon. Chem. = *Monatshefte für Chemie*
NBS Monogr. = *Monograph of the USA National Bureau of Standards*
N. Jb. Min. Abh. = *Neues Jahrbuch für Mineralogie, Abhandlungen*
N. Jb. Min. Mh. = *Neues Jahrbuch für Mineralogie, Monatshefte*
occ. = occupancy
O'Keeffe & Hyde = M. O'Keeffe and B.G. Hyde (1996) *Crystal structures I. Patterns and symmetry*. Miner. Soc. America, Washington.
part. occ. = partial occupancy
Pearson, 1967 = W.B. Pearson (1967) *Handbook of lattice spacings and structures of metals*. Pergamon Press, New York.
Pearson, 1972 = W.B. Pearson (1972) *The crystal chemistry and physics of metals and alloys*. Wiley Intersciences, New York.
Phys. Chem. Min. = *Physics and Chemistry of Minerals*

Pov. = A.S. Povarennykh (1972) *Crystal Chemical Classification of Minerals* (English translation, Plenum Press, New York)

Rhomb. desc. = Rhombohedral description

Ross = V. Ross (1958) Geochemistry, crystal structure and mineralogy of sulphides. *Economic Geology*, *52*, 755–774

RRW = W.L. Roberts, G.R. Rapp, Jr., and J. Weber (1974) *Encyclopedia of minerals*. Van Nostrand, New York.

Sov. Phys. Cryst. = *Soviet Physics Crystallography*

SB = *Strukturbericht*

SR = *Structure Reports*

Str. Tab. = H. Strunz (1982) *Mineralogische tabellen*. Akademische Verlagsgesellschaft, Leipzig

TMPM = *Tschemaks Mineralogische und Petrographische Mitteilungen*

Wells, 1962 = A.F. Wells (1962) *Structural inorganic chemistry*. Clarendon Press, Oxford

Wells, 1984 = A.F. Wells (1984) *Structural inorganic chemistry, Fifth edition*, Clarendon Press, Oxford

Wyckoff = R.W.G. Wyckoff, *Crystal Structures*, vol. 1, 1963; vol. 2, 1964; vol. 3, 1965; vol. 4, 1968

Zeit. Krist. = *Zeitschrift für Kristallographie*

ZVMO = *Zapiski Vsesoyuznogo Mineralogischeskogo Obshchestva*

Mineral Index

Subject Index

Author Index

* Referred to in the systematic tables

Solid Earth Sciences Library

1. E.I. Galperin: *The Polarization Method of Seismic Exploration.* 1983 ISBN 90-277-1555-6
2. S. Hastenrath: *The Glaciers of Equatorial East Africa.* 1984 ISBN 90-277-1572-6
3. R.K. Verma: *Gravity Field, Seismicity and Tectonics of the Indian Peninsula and the Himalayas.* 1985 ISBN 90-277-1864-4
4. R. Haenel, L. Rybach and L. Stegena (eds.): *Handbook of Terrestrial Heat-Flow Density Determination.* With Guidelines and Recommendations of the International Heat-Flow Commission. 1988 ISBN 90-277-2589-6
5. E. Roth and B. Poty (eds.): *Nuclear Methods of Dating.* 1989 ISBN 0-7923-0188-9
6. G. Wagner and P. Van den haute: *Fission-Track Dating.* 1992 ISBN 0-7923-1624-X
7. J. Lima-de-Faria: *Structural Mineralogy.* An Introduction. 1994 ISBN 0-7923-2821-3
8. I.M. Varentsov: *Manganese Ores of Supergene Zone: Geochemistry of Formation.* 1996 ISBN 0-7923-3906-1
9. A.A. Meyerhoff, I. Taner, A.E.L. Morris, W.B. Agocs, M. Kamen-Kaye, M.I. Bhat, N.C. Smoot and Dong R. Choi; D. Meyerhoff Hull (ed.): *Surge Tectonics: A New Hypothesis of Global Geodynamics.* 1996 ISBN 0-7923-4156-2
10. P. Van den haute and F. De Corte (eds.): *Advances in Fission-Track Geochronology.* A selection of papers presented at the International Workshop on Fission-Track Dating, Ghent, Belgium (1996). 1998 ISBN 0-7923-4904-0
11. J. Lima-de-Faria: *Structural Classification of Minerals.* Volume 1: Minerals with A, A_mB_n and ApBqCr General Chemical Formulas. 2001 Set: 0-7923-6893-2; ISBN 0-7923-6892-4

Kluwer Academic Publishers – Dordrecht / Boston / London

MIX
Papier aus verantwortungsvollen Quellen
Paper from responsible sources
FSC® C105338

If you have any concerns about our products,
you can contact us on
ProductSafety@springernature.com

In case Publisher is established outside the EU,
the EU authorized representative is:
Springer Nature Customer Service Center GmbH
Europaplatz 3, 69115 Heidelberg, Germany

Printed by Libri Plureos GmbH
in Hamburg, Germany